Fields Institute Monographs

VOLUME 31

For further volumes:
http://www.springer.com/series/10502

Fields Institute Monographs

VOLUME 31

The Fields Institute is a centre for research in the mathematical sciences, located in Toronto, Canada. The Institutes mission is to advance global mathematical activity in the areas of research, education and innovation. The Fields Institute is supported by the Ontario Ministry of Training, Colleges and Universities, the Natural Sciences and Engineering Research Council of Canada, and seven Principal Sponsoring Universities in Ontario (Carleton, McMaster, Ottawa, Toronto, Waterloo, Western and York), as well as by a growing list of Affiliate Universities in Canada, the U.S. and Europe, and several commercial and industrial partners.

For further volumes:
http://www.springer.com/series/10502

Javad Mashreghi

Derivatives of Inner Functions

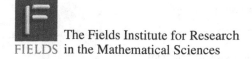
The Fields Institute for Research
in the Mathematical Sciences

Javad Mashreghi
Département de mathématiques et de
    statistique
Université Laval
Québec, QC
Canada

ISSN 1069-5273   ISSN 2194-3079 (electronic)
ISBN 978-1-4899-8941-3   ISBN 978-1-4614-5611-7 (eBook)
DOI 10.1007/978-1-4614-5611-7
Springer New York Heidelberg Dordrecht London

Mathematics Subject Classification (2010): 30D50, 30D40, 30D55, 30E20, 30E25, 32A36

Cover illustration: Drawing of J.C. Fields by Keith Yeomans

Printed on acid-free paper

Springer is part of Springer Science+Business Media (www.springer.com)

Ouvrez une école, vous fermerez une prison

Victor Hugo

It was in 1934 that Pahlavi High School was established in my hometown, Kashan. In 1946, the school moved to a new building constructed over a rather vast area and featuring an awesome architectural design. The school was renamed Imam Khomeini High School after the 1979 revolution. Over the years, numerous bright minds were trained in the stimulating environment of this school. Before long, the impressive school building had become a reminder of all the great intellectuals who had either studied or taught there. To the utter regret of the latter, however, the building was completely demolished in 1995, only to give way to the current, incomplete one. I dedicate this monograph to all the caring and respected men, teachers and employees alike, who kept the flame of education alight for many years in this institute.

مدرسه ای باز کنید، زندانی را خواهید بست.
ویکتور هوگو

دبیرستان پهلوی در شهر من کاشان به سال ۱۳۱۴ تاسیس و در سال ۱۳۲۶ در مکانی نسبتأ وسیع و با معماری بسیار زیبا و خاطره انگیز بنا شد. بعد از انقلاب ۱۳۵۷، نام اولیه دبیرستان به امام خمینی تغییر یافت. این مجموعه در طول سالیان متمادی نخبگان زیادی را در دامن پر مهر خود پرورش داد و بنای با شکوهش یاد و خاطره‌ی مردان بزرگ و گران‌قدری را که در آن تدریس و یا تحصیل نموده بودند زنده نگاه می‌داشت. با کمال تأسف در سال ۱۳۷۴، این بنا به طور کامل تخریب گردید و بنای جدید و ناتمام فعلی به جای آن ساخته شد. این کتاب را به دبیران دلسوز و کارمندان شریفی که مشعل آموزش را در این مرکز برافروخته داشته و به دوش کشیدند تقدیم می‌دارم.

Painting by Mrs. Najmeh Firoozpour

Preface

Infinite Blaschke products were introduced by W. Blaschke in 1915 [9]. In 1929, R. Nevanlinna introduced the class of bounded analytic functions with almost everywhere unimodular boundary values [35]. However, the term *inner function* was coined much later by A. Beurling in his seminal work on the invariant subspaces of the shift operator [8]. The first extensive studies of the properties of inner functions were made by O. Frostman [22], W. Seidel [43] and F. Riesz [40]. Their efforts to understand the zeros and boundary behavior of bounded analytic functions led to the celebrated canonical factorization theorem. The special factorization that we need says that each inner function is the product of a *Blaschke product* and a zero free inner function, the so called *singular inner function*, which is generated by a singular measure residing on the unit circle. Roughly speaking, we can say that the Blaschke product is formed with the zeros of an inner function inside the open unit disc, and the singular part stems from its zeros on the boundary.

In July 2011, E. Fricain and I organized a conference on *Blaschke products and their applications* in the Fields Institute (Toronto). There were several interesting talks about the boundary behavior of inner functions, in particular Blaschke products, and their derivatives. I felt the need to gather some classical results in a short monograph for graduate students and as a handy reference for experts. However, the literature is very vast and it is a difficult task to choose among various important results. For example, the book of P. Colwell [16] can provide a panoramic picture of this subject. Hence, I restricted myself just to the integral means of the derivatives and, even for this narrow subject, I was very selective.

The Fields Institute exclusively supported our conference on Blaschke products, and its direction constantly helped us for the production of the proceedings and this monograph. In particular, I owe profound thanks to Carl Riehm, the Managing Editor of Publications, for his care, guidance, and enthusiastic support. Last but not the least, I would like to deeply thank

Joseph Cima (University of North Carolina), Ian Graham (University of Toronto), and Armen Edigarian (Jagiellonian University) who kindly read the manuscript and made many valuable suggestions. Their remarks enormously improved the quality of text.

Montreal, QC   Javad Mashreghi

Contents

Chapter 1
Inner Functions

The theory of Hardy spaces is a well established part of analytic function theory. Inner functions constitute a special family in this category. Therefore, it is natural to start with several topics on Hardy spaces and apply them in our discussions. However, we are not in a position to study this theory in detail and we assume that our readers have an elementary familiarity with this subject. In this chapter, we briefly mention, mostly without proof, the main theorems that we need in the study of inner functions. For a detailed study of this topic, we refer to [33].

1.1 The Poisson Integral of a Measure

Let μ be a complex Borel measure on the unit circle $\mathbb{T}$. Then the Poisson integral of μ on the open unit disc $\mathbb{D}$ is defined by the formula

$$P_\mu(z) = \int_\mathbb{T} \frac{1 - |z|^2}{|z - \zeta|^2} \, d\mu(\zeta), \qquad (z \in \mathbb{D}).$$

If $d\mu(e^{i\theta}) = u(e^{i\theta}) \, d\theta/2\pi$, where $u \in L^1(\mathbb{T})$, instead of P_μ we write P_u. It is easy to verify that $h = P_\mu$ is a harmonic function on $\mathbb{D}$. Moreover, using Fubini's theorem and the identity

$$\frac{1}{2\pi} \int_0^{2\pi} \frac{1 - |z|^2}{|z - e^{i\theta}|^2} \, d\theta = 1, \qquad (z \in \mathbb{D}), \tag{1.1}$$

we see that

$$\frac{1}{2\pi} \int_0^{2\pi} |h(re^{i\theta})| \, d\theta \le \int_\mathbb{T} \left(\frac{1}{2\pi} \int_0^{2\pi} \frac{1 - r^2}{|re^{i\theta} - \zeta|^2} \, d\theta \right) d|\mu|(\zeta) = \|\mu\|,$$

J. Mashreghi, *Derivatives of Inner Functions*, Fields Institute Monographs 31, DOI 10.1007/978-1-4614-5611-7_1, © Springer Science+Business Media New York 2013

where $\|\mu\|$ is the total variation of the measure μ on $\mathbb{T}$. Hence, h fulfills the growth restriction

$$\sup_{0 \le r < 1} \int_0^{2\pi} |h(re^{i\theta})| \, d\theta < \infty. \tag{1.2}$$

Hence, the Poisson integral of a Borel measure on $\mathbb{T}$ is a harmonic function on $\mathbb{D}$ which satisfies (1.2). As a matter of fact, the converse to this assertion is also true and we have the following complete characterization.

Theorem 1.1 (Plessner [36]) *Let h be a function defined on $\mathbb{D}$. Then the following assertions are equivalent:*

(i) h is a harmonic function on $\mathbb{D}$ which satisfies the condition (1.2);
(ii) there exists a (unique) Borel measure μ on $\mathbb{T}$ such that $h = P_\mu$.

As a special case, if μ is positive, then $h = P_\mu$ is a positive harmonic function on $\mathbb{D}$ which satisfies (1.2). And if h is a given positive harmonic function, then, by the mean value property, it satisfies

$$\int_0^{2\pi} |h(re^{i\theta})| \, d\theta = \int_0^{2\pi} h(re^{i\theta}) \, d\theta = 2\pi h(0), \qquad (0 \le r < 1).$$

Therefore, in this case, Theorem 1.1 is rewritten as follows.

Corollary 1.2 (Herglotz [28]) *Let h be a function defined on $\mathbb{D}$. Then the following assertions are equivalent:*

(i) h is a positive harmonic function on $\mathbb{D}$;
(ii) there exists a (unique) positive Borel measure μ on $\mathbb{T}$ such that $h = P_\mu$.

The following celebrated result of Fatou provides a sufficient condition for the existence of radial limits of P_μ.

Theorem 1.3 (Fatou [20]) *Let μ be a complex Borel measure on $\mathbb{T}$. Suppose that at $e^{i\theta} \in \mathbb{T}$ the symmetric derivative*

$$\mu'(e^{i\theta}) = \lim_{t \to 0} \frac{\mu(\{e^{is} : s \in (\theta - t, \theta + t)\})}{2t}$$

exists. Then

$$\lim_{r \to 1} P_\mu(re^{i\theta}) = 2\pi \, \mu'(e^{i\theta}).$$

Proof. Without loss of generality, assume that $\theta = 0$. Put

$$\mathfrak{U}(x) = \mu(\{e^{is} : s \in [-\pi, x)\}), \qquad x \in [-\pi, \pi).$$

Then integration by parts gives

$$P_\mu(r) = \int_{\mathbb{T}} \frac{1 - r^2}{1 + r^2 - 2r\cos t} \, d\mu(e^{it})$$

$$= \left\{ \frac{1 - r^2}{1 + r^2 - 2r\cos t} \, \mathfrak{U}(t) \right\} \Big|_{-\pi}^{\pi} - \int_{-\pi}^{\pi} \frac{\partial}{\partial t} \left\{ \frac{1 - r^2}{1 + r^2 - 2r\cos t} \right\} \mathfrak{U}(t) \, dt$$

$$= \frac{1 - r}{1 + r} \mathfrak{U}(\pi) + \int_{-\pi}^{\pi} \frac{(1 - r^2) \, 2r \sin t}{(1 + r^2 - 2r\cos t)^2} \, \mathfrak{U}(t) \, dt$$

$$= \frac{1 - r}{1 + r} \mathfrak{U}(\pi) + \frac{2r}{1 + r} \int_{-\pi}^{\pi} \frac{(1 + r)^2 \, (1 - r) \, t \sin t}{(1 + r^2 - 2r\cos t)^2} \times \frac{\mathfrak{U}(t) - \mathfrak{U}(-t)}{2t} \, dt.$$

Let

$$\phi(t) = \frac{\mathfrak{U}(t) - \mathfrak{U}(-t)}{2t} - \mu'(1) = \frac{1}{2t} \int_{-t}^{t} d\mu(e^{is}) - \mu'(1), \qquad (-\pi \le t \le \pi),$$

and note that, by assumption,

$$\lim_{t \to 0} \phi(t) = 0. \tag{1.3}$$

Let

$$F_r(t) = \frac{(1 + r)^2 \, (1 - r) \, t \sin t}{(1 + r^2 - 2r\cos t)^2}, \qquad (0 \le r < 1, \; -\pi \le t \le \pi).$$

This function satisfies the following properties:

(i) $F_r \ge 0$;
(ii)

$$\frac{1}{2\pi} \int_{-\pi}^{\pi} F_r(t) \, dt = 1;$$

(iii) for each fixed $0 < \delta < \pi$, we have

$$\lim_{r \to 0} \left(\sup_{\delta < |t| \le \pi} F_r(t) \right) = 0.$$

In technical language, F_r is a positive approximate identity on $[-\pi, \pi]$. Using the new notations, we have

$$\lim_{r \to 1} P_\mu(r) = \lim_{r \to 1} \int_{-\pi}^{\pi} F_r(t) \left(\phi(t) + \mu'(1) \right) dt = 2\pi\mu'(1) + \lim_{r \to 1} \int_{-\pi}^{\pi} F_r(t) \, \phi(t) \, dt.$$

By (1.3), given $\varepsilon > 0$, there is δ such that $|\phi(t)| < \varepsilon$, whenever $|t| < \delta$. Without loss of generality, assume that $\delta < \pi$. Then we have

$$\left| \int_{-\pi}^{\pi} F_r(t)\, \phi(t)\, dt \right| \leq \int_{-\delta}^{\delta} F_r(t)\, |\phi(t)|\, dt + \int_{\delta < |t| \leq \pi} F_r(t)\, |\phi(t)|\, dt$$

$$\leq \varepsilon \int_{-\delta}^{\delta} F_r(t)\, dt + \left(\max_{-\pi \leq t \leq \pi} |\phi(t)| \right) \int_{\delta < |t| \leq \pi} F_r(t)\, dt$$

$$\leq 2\pi\varepsilon + \pi \left(\max_{-\pi \leq t \leq \pi} |\phi(t)| \right) \left(\sup_{\delta < |t| \leq \pi} F_r(t) \right).$$

Therefore, for each $\varepsilon > 0$,

$$\limsup_{r \to 1} \left| \int_{-\pi}^{\pi} F_r(t)\, \phi(t)\, dt \right| \leq 2\pi\varepsilon.$$

This fact ensures that

$$\lim_{r \to 1} P_\mu(r) = 2\pi\mu'(1).$$

By *Lebesgue's decomposition theorem*, for each complex Borel measure μ, there are a function $u \in L^1(\mathbb{T})$ and a complex singular Borel measure σ such that

$$d\mu(e^{i\theta}) = u(e^{i\theta})\, d\theta/2\pi + d\sigma(e^{i\theta}).$$

Moreover, for almost all $e^{i\theta} \in \mathbb{T}$,

$$\mu'(e^{i\theta}) = \lim_{t \to 0} \frac{\mu\big(\{ e^{is} : s \in (\theta - t, \theta + t) \} \big)}{2t} = \frac{u(e^{i\theta})}{2\pi}.$$

Hence, we immediately obtain the following two results. First, if $\mu = \sigma$ is a complex singular Borel measure on $\mathbb{T}$, then

$$\lim_{r \to 1} P_\sigma(re^{i\theta}) = 0 \tag{1.4}$$

for almost all $e^{i\theta} \in \mathbb{T}$. Second, if $d\mu = u\, d\theta/2\pi$ is absolutely continuous, then

$$\lim_{r \to 1} P_u(re^{i\theta}) = u(e^{i\theta}) \tag{1.5}$$

for almost all $e^{i\theta} \in \mathbb{T}$.

The following variant of Fatou's theorem will also be needed. Since F_r is a positive approximate identity, the proof of Theorem 1.3, with slight modification, works in this case too.

Theorem 1.4 *Let μ be a finite positive Borel measure on $\mathbb{T}$, and let $e^{i\theta} \in \mathbb{T}$ be such that*

$$\mu'(e^{i\theta}) = \lim_{t \to 0} \frac{\mu\big(\{ e^{is} : s \in (\theta - t, \theta + t) \} \big)}{2t} = \infty.$$

Then

$$\lim_{r \to 1} P_\mu(re^{i\theta}) = \infty.$$

Proof. Without loss of generality, assume that $\theta = 0$. We use the same notations as in the proof of Theorem 1.3. We have

$$\lim_{r \to 1} P_\mu(r) = \lim_{r \to 1} \int_{-\pi}^{\pi} F_r(t)\,\phi(t)\,dt,$$

except that in this case

$$\phi(t) = \frac{\mathfrak{U}(t) - \mathfrak{U}(-t)}{2t} = \frac{1}{2t} \int_{-t}^{t} d\mu(e^{is}) \geq 0, \qquad (-\pi \leq t \leq \pi),$$

and

$$\lim_{t \to 0} \phi(t) = \infty.$$

Hence, given $M > 0$, there is δ such that $\phi(t) \geq M$, whenever $|t| < \delta$. Without loss of generality, assume that $\delta < \pi$. Then we have

$$\int_{-\pi}^{\pi} F_r(t)\,\phi(t)\,dt \geq \int_{-\delta}^{\delta} F_r(t)\,\phi(t)\,dt$$

$$\geq M \int_{-\delta}^{\delta} F_r(t)\,dt$$

$$= M \left(2\pi - \int_{\delta < |t| \leq \pi} F_r(t)\,dt \right)$$

$$\geq 2\pi M - M \left(\sup_{\delta < |t| \leq \pi} F_r(t) \right).$$

Therefore, for each $M > 0$,

$$\liminf_{r \to 1} \int_{-\pi}^{\pi} F_r(t)\,\phi(t)\,dt \geq 2\pi M.$$

This fact ensures that

$$\lim_{r \to 1} P_\mu(r) = \infty.$$

The set

$$S_C(e^{i\theta}) = \{ z \in \mathbb{D} : |z - e^{i\theta}| \leq C\,(1 - |z|) \},$$

where $C > 1$ is a constant, is called a Stolz domain. This domain looks like an angle with vertex at $e^{i\theta}$, and the bigger C yields the wider angle. A function f defined on $\mathbb{D}$ has a non-tangential limit at $e^{i\theta}$ if, for each fixed constant C, the limit

$$\lim_{\substack{z \longrightarrow e^{i\theta} \\ z \in S_C(e^{i\theta})}} f(z)$$

exists and its value is independent of C. In this situation, we write

$$\lim_{\substack{z \longrightarrow e^{i\theta} \\ \vartriangleleft}} f(z).$$

In all preceding results, the radial limit can be replaced by a nontangential limit. The proofs become slightly more complicated. However, the same techniques work.

In some applications, we will also need the approach region

$$S_{C,\delta}(e^{i\theta}) = \left\{ z \in \mathbb{D} : |e^{i\theta} - z| \le C(1 - |z|)^{\delta} \right\},$$

where $C \ge 1$ and $0 < \delta \le 1$. This is a generalized Stolz domain anchored at $e^{i\theta} \in \mathbb{T}$. The smaller δ, the more tangential $S_{C,\delta}$ is to the boundary at $e^{i\theta}$.

1.2 The Hardy Space $H^p(\mathbb{D})$

Let f be an analytic function on the open unit disc $\mathbb{D}$. Define

$$\| f \|_p = \sup_{0 \le r < 1} \| f_r \|_p = \sup_{0 \le r < 1} \left(\frac{1}{2\pi} \int_0^{2\pi} |f(re^{i\theta})|^p \, d\theta \right)^{\frac{1}{p}},$$

if $p \in (0, \infty)$, and

$$\| f \|_\infty = \sup_{z \in \mathbb{D}} | f(z) |.$$

Then the *Hardy space* $H^p(\mathbb{D})$ is the family of all analytic functions f which satisfy the growth restriction $\|f\|_p < \infty$. The Hardy spaces H^1, H^2 and H^∞ will frequently appear in our discussion. A simple application of Hölder's inequality shows that

$$H^\infty(\mathbb{D}) \subset H^q(\mathbb{D}) \subset H^p(\mathbb{D})$$

for $0 < p < q < \infty$. In particular, we have $H^\infty \subset H^2 \subset H^1$.

According to (1.5), for each $f \in H^p(\mathbb{D})$, $0 < p \le \infty$, the non-tangential limit

$$f^*(e^{i\theta}) = \lim_{\substack{z \longrightarrow e^{i\theta} \\ \vartriangleleft}} f(z) \tag{1.6}$$

exists for almost all $e^{i\theta} \in \mathbb{T}$. Moreover, $f^* \in L^p(\mathbb{T})$ and $\|f^*\|_{L^p(\mathbb{T})} = \|f\|_{H^p(\mathbb{D})}$, $0 < p \le \infty$, and

$$\lim_{r \to 1} \| f_r - f^* \|_p = 0, \qquad (0 < p < \infty). \tag{1.7}$$

This result establishes a norm preserving correspondence between $H^p(\mathbb{D})$ and a closed subspaces of $L^p(\mathbb{T})$, which we denote by $H^p(\mathbb{T})$. Hence, without facing any essential difficulty, we will also write f for f^*. For any nonzero $f \in H^p$, we have

$$\log|f| \in L^1(\mathbb{T}) \tag{1.8}$$

and

$$\log|f(re^{i\theta})| \leq \frac{1}{2\pi} \int_0^{2\pi} \frac{1-r^2}{1+r^2-2r\cos(\theta-t)} \log|f(e^{it})| \, dt. \tag{1.9}$$

If $1 \leq p \leq \infty$, $H^p(\mathbb{T})$ has the equivalent characterization

$$H^p(\mathbb{T}) = \{f \in L^p(\mathbb{T}) : \hat{f}(n) = 0, \ n \leq -1\},$$

where

$$\hat{f}(n) = \frac{1}{2\pi} \int_0^{2\pi} f(e^{it}) \, e^{-int} \, dt, \qquad (n \in \mathbb{Z}),$$

is the n-th Fourier coefficient of f. We also have

$$f(z) = \frac{1}{2\pi} \int_0^{2\pi} \frac{1-|z|^2}{|z-e^{it}|^2} f(e^{it}) \, dt, \qquad (z \in \mathbb{D}), \tag{1.10}$$

or equivalently

$$f(re^{i\theta}) = \frac{1}{2\pi} \int_0^{2\pi} \frac{1-r^2}{1+r^2-2r\cos(\theta-t)} f(e^{it}) \, dt, \qquad (re^{i\theta} \in \mathbb{D}). \tag{1.11}$$

In other words, f is the Poisson integral of its boundary values.

A function $f \in L^\infty(\mathbb{T})$ is called *unimodular* if

$$|f(e^{it})| = 1$$

for almost all $e^{it} \in \mathbb{T}$. A function $f \in H^\infty(\mathbb{D})$ is called *inner* if it is unimodular on the boundary. Note that if an analytic function on $\mathbb{D}$ has unimodular boundary values almost everywhere on $\mathbb{T}$, we cannot deduce that it is an inner function. For example, the function

$$f(z) = \exp\left(\frac{1+z}{1-z}\right), \qquad (z \in \mathbb{D}),$$

satisfies

$$f(e^{i\theta}) = \exp\left(i\cot(\theta/2)\right)$$

for all $e^{i\theta} \in \mathbb{T} \setminus \{1\}$. However, f is not an inner function, since

$$f(r) = \exp\left(\frac{1+r}{1-r}\right) \longrightarrow \infty$$

as $r \longrightarrow 1$.

1.3 Two Classes of Inner Functions

This section is very important for the whole subject. We introduce two classes
of inner functions: Blaschke products and singular inner functions. Then we
will see that any arbitrary inner function has a unique decomposition as the
product of two such inner functions.

Let $\{z_n\}_{n\geqslant 1}$ be a sequence of nonzero complex numbers inside the open
unit disc $\mathbb{D}$ with $\lim_{n\to\infty}|z_n| = 1$. The sequence is indexed such that $0 <
|z_1| \leqslant |z_2| \leqslant \cdots$. It is worthwhile to mention that each number z_n can repeat
a finite number of times. Let

$$\Omega = \Omega_{(z_n)} = \mathbb{C} \setminus \left\{ \frac{1}{\bar{z}_n} : n \geq 1 \right\}^{cl}.$$

The notion A^{cl} represents the topological closure of A in the complex plane.
Since $\lim_{n\to\infty}|z_n| = 1$, all accumulation points of $\{1/\bar{z}_n : n \geq 1\}$ are on the
unite circle $\mathbb{T}$. Clearly, this set has at least one accumulation point. On the
one hand, the set might have just one extreme point, e.g. when all the points
are on a radius. On the other hand, we can easily construct a sequence which
accumulates at all points of $\mathbb{T}$, e.g. $e^{i\theta_n}/n^2$ where $(\theta_n)_{n\geq 1}$ is an enumeration
of the rational numbers. Thus the domain Ω always contains open unit disc
$\mathbb{D}$, and $\mathbb{D}_e \setminus \{1/\bar{z}_n : n \geq 1\}$, where $\mathbb{D}_e = \{z : |z| > 1\}$. Moreover, it may also
contain some open arcs of $\mathbb{T}$.

Fix $m \in \mathbb{N}$, $\beta \in \mathbb{R}$, and put

$$B_k(z) = e^{i\beta} z^m \prod_{n=1}^{k} \frac{|z_n|}{z_n} \frac{z_n - z}{1 - \bar{z}_n z}. \tag{1.12}$$

Under a certain condition on the rate of growth of $(|z_n|)_{n\geq 1}$, the partial
products B_k, $k \geq 1$, converge uniformly on compact subsets of Ω to an
analytic function, which we denote by

$$B(z) = e^{i\beta} z^m \prod_{n=1}^{\infty} \frac{|z_n|}{z_n} \frac{z_n - z}{1 - \bar{z}_n z}. \tag{1.13}$$

The function B is called an *infinite Blaschke product* for the open unit disc $\mathbb{D}$.

Theorem 1.5 (Blaschke [9]) *Let the points $z_n \in \mathbb{D} \setminus \{0\}$, $n \geq 1$, be such
that $\lim_{n\to\infty}|z_n| = 1$. Then a necessary and sufficient condition for the uni-
form convergence of B_k, given by (1.12), on compact subsets of*

$$\Omega = \mathbb{C} \setminus \left\{ \frac{1}{\bar{z}_n} : n \geqslant 1 \right\}^{cl}$$

to a nonzero analytic function is that

$$\sum_{n=1}^{\infty}(1-|z_n|) < \infty.$$

Remark. There is no restriction on $\arg z_n$.

Proof. Suppose that B_k is uniformly convergent to B on compact subsets of Ω. Without loss of generality, we can assume that $B(0) \neq 0$. Since otherwise, we can consider the partial products $B_k(z)/z^m$, which uniformly converge to $B(z)/z^m$ on compact subsets of Ω. The simple inequality

$$t \leqslant e^{t-1}, \quad 0 \leqslant t \leqslant 1,$$

implies

$$|B(0)| = \prod_{n=1}^{\infty}|z_n| \leqslant \exp\left(-\sum_{n=1}^{\infty}(1-|z_n|)\right).$$

Since $B(0) \neq 0$, then

$$\sum_{n=1}^{\infty}(1-|z_n|) < \infty.$$

Now, for the reverse implication, suppose that the last inequality holds. The identity

$$\frac{|z_n|}{z_n}\frac{z_n-z}{1-\bar{z}_n z} = 1 - (1-|z_n|)\frac{z_n+|z_n|z}{z_n(1-\bar{z}_n z)} \tag{1.14}$$

implies

$$\left|1 - \frac{|z_n|}{z_n}\frac{z_n-z}{1-\bar{z}_n z}\right| \leq (1-|z_n|)\frac{1+|z|}{|1-\bar{z}_n z|}.$$

On compact subsets of Ω, we have

$$\frac{1+|z|}{|1-\bar{z}_n z|} \leq \frac{1+|z|}{|z_n||z-1/\bar{z}_n|} \leq \frac{1+|z|}{|z_1|\,\mathrm{dist}(z,\partial\Omega)} \leq C,$$

where C is a constant independent of n. However, C depends on the choice of compact set, but this dependence is harmless. Therefore, for all z in a compact subset of Ω,

$$\left|1 - \frac{|z_n|}{z_n}\frac{z_n-z}{1-\bar{z}_n z}\right| \leqslant C(1-|z_n|), \qquad (n \geq 1).$$

This inequality establishes the uniform and absolute convergence of the partial products B_k on compact subsets of Ω.

A sequence $\{z_n\}$ of complex numbers in the open unit disc satisfying the condition

$$\sum_{n=1}^{\infty} \left(1 - |z_n|\right) < \infty. \tag{1.15}$$

is called a *Blaschke sequence*. The growth restriction (1.15) is also known as the *Blaschke condition*. Under this condition, the partial products B_k converge uniformly on any compact subset of $\mathbb{D}$ to the infinite Blaschke product B. Since $|B_k(z)| \leq 1$ on $\mathbb{D}$, we necessarily have

$$|B(z)| \leq 1, \qquad (z \in \mathbb{D}). \tag{1.16}$$

Hence, a Blaschke product B, restricted to $\mathbb{D}$, is in $H^{\infty}(\mathbb{D})$. Thus, by (1.6), for almost all $e^{i\theta} \in \mathbb{T}$,

$$B(e^{i\theta}) = \lim_{\substack{z \to e^{i\theta} \\ \triangleleft}} B(z)$$

exists and satisfies the norm identity

$$\| B \|_{H^{\infty}(\mathbb{D})} = \| B \|_{L^{\infty}(\mathbb{T})}.$$

Moreover, by (1.11),

$$B(re^{i\theta}) = \frac{1}{2\pi} \int_0^{2\pi} \frac{1 - r^2}{1 + r^2 - 2r\cos(\theta - t)} B(e^{it}) \, dt, \qquad (re^{i\theta} \in \mathbb{D}).$$

If B is a finite Blaschke product, then $B(e^{i\theta})$ is well defined analytic function on $\mathbb{T}$. But, if B is an infinite Blaschke product, then $B(e^{i\theta})$ is defined almost everywhere on $\mathbb{T}$. According to Theorem 1.5, an infinite Blaschke product is analytic and unimodular at $e^{i\theta} \in \mathbb{T}$ if and only if $e^{i\theta}$ is not an accumulation point of its zeros. But, if $e^{i\theta} \in \mathbb{T}$ is an accumulation point of the zeros of B, then B is not even continuous there. Nevertheless, $B(e^{i\theta})$ is still unimodular almost everywhere on $\mathbb{T}$. This result is of special interest when the zeros of B accumulate on a subset of $\mathbb{T}$ with a positive measure. In particular, the zeros of B may accumulate at all points of $\mathbb{T}$.

Lemma 1.6 *Let B be a Blaschke product for the unit disc. Then*

$$\lim_{r \to 1} \frac{1}{2\pi} \int_0^{2\pi} \log|B(re^{i\theta})| \, d\theta = 0.$$

Proof. By (1.16),

$$\frac{1}{2\pi} \int_0^{2\pi} \log|B(re^{i\theta})| \, d\theta \leq 0 \tag{1.17}$$

for all r, $0 \leq r < 1$.

Without loss of generality, we assume that $B(0) \neq 0$. Since otherwise, we can divide B by z^m, where m is the order of B at the origin, and this modification does not change the limit that we want to evaluate. Now, by Jensen's formula,

$$\log |B(0)| = \sum_{|z_n|<r} \log \left(\frac{|z_n|}{r} \right) + \frac{1}{2\pi} \int_0^{2\pi} \log |B(re^{i\theta})|\, d\theta$$

for all r, $0 < r < 1$. Since $B(0) = \prod_{n=1}^{\infty} |z_n|$, we thus obtain

$$\frac{1}{2\pi} \int_0^{2\pi} \log |B(re^{i\theta})|\, d\theta = \sum_{|z_n|<r} \log \left(\frac{r}{|z_n|} \right) - \sum_{n=1}^{\infty} \log \left(\frac{1}{|z_n|} \right).$$

At this point we can appeal to the monotone convergence theorem and finish the proof. However, a direct proof is also possible. Given $\varepsilon > 0$, choose N so large that

$$\sum_{n=N+1}^{\infty} \log \left(\frac{1}{|z_n|} \right) < \varepsilon.$$

Then, for $r > |z_N|$, we have

$$\frac{1}{2\pi} \int_0^{2\pi} \log |B(re^{i\theta})|\, d\theta \geqslant \sum_{n=1}^{N} \log \left(\frac{r}{|z_n|} \right) - \sum_{n=1}^{N} \log \left(\frac{1}{|z_n|} \right) - \varepsilon.$$

Therefore,

$$\liminf_{r \to 1} \frac{1}{2\pi} \int_0^{2\pi} \log |B(re^{i\theta})|\, d\theta \geqslant -\varepsilon,$$

and, since ε is an arbitrary positive number,

$$\liminf_{r \to 1} \frac{1}{2\pi} \int_0^{2\pi} \log |B(re^{i\theta})|\, d\theta \geqslant 0.$$

Finally, (1.17) and the last inequality imply that the limit exists and is zero.

The property described in Lemma 1.6 has several applications. In fact, we will see that it is a characterization of Blaschke products among inner functions. In the first step, let us see how it implies that B is an inner function.

Theorem 1.7 (Riesz [40]) *Let B be a Blaschke product for the unit disc. Then*

$$|B(e^{i\theta})| = 1$$

for almost all $e^{i\theta} \in \mathbb{T}$.

Proof. By (1.16), we have $|B(e^{i\theta})| \leq 1$ for almost all $e^{i\theta} \in \mathbb{T}$. Moreover, by Fatou's lemma and Lemma 1.6,

$$\frac{1}{2\pi} \int_0^{2\pi} \log \left| \frac{1}{B(e^{i\theta})} \right| d\theta \leq 0. \tag{1.18}$$

Since $\log\big|1/B(e^{i\theta})\big| \geqslant 0$, and (1.18) holds, we must have $|B(e^{i\theta})| = 1$ for almost all $e^{i\theta} \in \mathbb{T}$.

We now introduce the singular inner functions. Let σ be a *positive singular Borel measure* on $\mathbb{T}$, and let

$$S(z) = S_\sigma(z) = \exp\left(-\int_{\mathbb{T}} \frac{e^{it} + z}{e^{it} - z}\, d\sigma(t) \right). \tag{1.19}$$

Clearly S is analytic on $\mathbb{D}$. Since

$$|S_\sigma(re^{i\theta})| = \exp\left(-\int_{\mathbb{T}} \frac{1 - r^2}{1 + r^2 - 2r\cos(\theta - t)}\, d\sigma(t) \right) \tag{1.20}$$

and σ is positive, then $|S(re^{i\theta})| \leq 1$ for all $re^{i\theta} \in \mathbb{D}$, i.e. $S \in H^\infty(\mathbb{D})$. Moreover, by (1.4),

$$\lim_{r \to 1} |S(re^{i\theta})| = 1$$

for almost all $e^{i\theta} \in \mathbb{T}$. In other words, S is an inner function. The function S is called a *singular inner* function. We end this section by describing a property which always holds for singular inner functions. But, finite Blaschke products never fulfil such a property. We will also construct infinite Blaschke products which does not show such a behavior. See Theorem 7.1.

Lemma 1.8 *Let σ be a positive singular Borel measure on $\mathbb{T}$, with $\sigma \neq 0$. Then there is at least one point $e^{i\theta} \in \mathbb{T}$ such that*

$$\lim_{r \to 1} S_\sigma(re^{i\theta}) = 0.$$

Proof. Since σ is singular with respect to Lebesgue measure

$$\mu'(e^{i\theta}) = \lim_{t \to 0} \frac{\mu\big(\{e^{is} : s \in (\theta - t, \theta + t)\}\big)}{2t} = \infty.$$

for all $e^{i\theta} \in E$, where $E \subset \mathbb{T}$ is such that $\sigma(\mathbb{T} \setminus E) = 0$. Since $\sigma \neq 0$, we surely have $E \neq \emptyset$. Hence, by Theorem 1.4, at all points of E, we have

$$\lim_{r \to 1} \int_{\mathbb{T}} \frac{1 - r^2}{1 + r^2 - 2r\cos(\theta - t)}\, d\sigma(t) = \infty,$$

which, by (1.20), implies $S_\sigma(re^{i\theta}) \longrightarrow 0$.

1.4 The Canonical Factorization

According to a celebrated result of Weierstrass, if $(z_n)_{n \geq 1}$ is a sequence in $\mathbb{D}$ which does not cluster at any point inside $\mathbb{D}$, or equivalently

$$\lim_{n\to\infty} |z_n| = 1,$$

then there is an analytic function f on $\mathbb{D}$, $f \not\equiv 0$, such that

$$f(z_n) = 0, \qquad (n \geq 1).$$

We even can choose f such that it has no other zeros. However, if we put some restrictions on the rate of growth of $|f|$, as we approach to the boundary, naturally it will cause some restriction on the rate of growth of zeros of f. The following result is a manifestation of this fact.

Lemma 1.9 *Let f be an analytic function on $\mathbb{D}$, $f \not\equiv 0$, and let $(z_n)_{n\geq 1}$ denote the sequence of its zeros in $\mathbb{D}$. Then*

$$\sum_{n=1}^{\infty} (1 - |z_n|) < \infty$$

if and only if

$$\sup_{0\leq r<1} \int_0^{2\pi} \log|f(re^{i\theta})| \, d\theta < \infty.$$

Remark. If f has a finite number of zeros on $\mathbb{D}$, then we replace $\sum_{n=1}^{\infty}$ by $\sum_{n=1}^{N}$.

Proof. Put

$$M = \sup_{0\leq r<1} \int_0^{2\pi} \log|f(re^{i\theta})| \, d\theta.$$

Without loss of generality assume that $f(0) \neq 0$. Then, by Jensen's formula,

$$\frac{1}{2\pi} \int_0^{2\pi} \log|f(re^{i\theta})| \, d\theta = \log|f(0)| + \sum_{|z_n|<r} \log \frac{r}{|z_n|}, \qquad (0 < r < 1).$$

Hence, either by the monotone convergence theorem or by direct verification as we did in Lemma 1.6,

$$M = \log|f(0)| + \sum_{n=1}^{\infty} \log \frac{1}{|z_n|}.$$

Therefore, $M < \infty$ if and only if

$$\sum_{n=1}^{\infty} \log|z_n| > -\infty,$$

which is equivalent to the Blaschke condition.

If $f \in H^p(\mathbb{D})$, $0 < p < \infty$ and $f \not\equiv 0$, then

$$\exp\left(\frac{1}{2\pi}\int_0^{2\pi} \log|f(re^{i\theta})|^p\, d\theta\right) \leq \int_0^{2\pi} |f(re^{i\theta})|^p\, d\theta$$
$$\leq \|f\|_p^p.$$

Thus,

$$\frac{1}{2\pi}\int_0^{2\pi} \log|f(re^{i\theta})|\, d\theta \leq \log\|f\|_p.$$

This inequality trivially holds for $p = \infty$. Lemma 1.9 now ensures that we can form a Blaschke product with zeros of f. F. Riesz discovered that we can extract these zeros in a special way.

Theorem 1.10 (Riesz [40]) *Let* $f \in H^p(\mathbb{D})$, $f \not\equiv 0$, *and let* B *be the Blaschke product formed with the zeros of* f *in* $\mathbb{D}$. *Then there is a zero free* $g \in H^p(\mathbb{D})$ *such that*
$$f = B\, g,$$

and

$$\|f\|_p = \|g\|_p.$$

Proof. Put $g = f/B$. Clearly g is a zero free analytic on $\mathbb{D}$ and, moreover, $\|f\|_p \leq \|g\|_p$.

First, suppose that $0 < p < \infty$. Fix $N \geq 1$ and $0 < r < 1$ such that there is no zeros of f on the circle $\{|z| = r\}$. Let B_N be the finite Blaschke product formed with the first N zeros of f. Then $g_N = f/B_N$ is an analytic function on $\mathbb{D}$, and

$$\frac{1}{2\pi}\int_0^{2\pi} |g_N(re^{i\theta})|^p\, d\theta \leq \frac{1}{2\pi}\int_0^{2\pi} |g_N(\rho e^{i\theta})|^p\, d\theta$$
$$\leq \frac{1}{\inf_\theta |B_N(\rho e^{i\theta})|^p}\left(\frac{1}{2\pi}\int_0^{2\pi} |f(\rho e^{i\theta})|^p\, d\theta\right)$$
$$\leq \frac{\|f\|_p^p}{\inf_\theta |B_N(\rho e^{i\theta})|^p}$$

for all ρ, with $r < \rho < 1$. Let $\rho \longrightarrow 1$. Since $|B_N|$ uniformly tends to 1, then

$$\frac{1}{2\pi}\int_0^{2\pi} |g_N(re^{i\theta})|^p\, d\theta \leq \|f\|_p^p.$$

Now, let $N \longrightarrow \infty$. On the circle $\{|z| = r\}$, B_N tends uniformly to B. Hence,

$$\frac{1}{2\pi}\int_0^{2\pi} |g(re^{i\theta})|^p\, d\theta \leq \|f\|_p^p.$$

Finally, let $r \longrightarrow 1$ through a sequence r_n such that there is no zeros of f on the circles $\{|z| = r_n\}$ to get $\|g\|_p \leq \|f\|_p$.

With some minor modifications, the proof also works for $p = \infty$.

Corollary 1.11 *Let $0 < p < \infty$, and let $f \in H^p(\mathbb{D})$, $f \not\equiv 0$. Let B be the Blaschke product formed with the zeros of f in $\mathbb{D}$. Then there is $g \in H^2(\mathbb{D})$, with no zeros on $\mathbb{D}$, such that*

$$f = B \, g^{2/p}.$$

Moreover,

$$\|f\|_p^p = \|g\|_2^2.$$

Proof. By Theorem 1.10, f has the decomposition $f = Bh$, where h is a zero free element of $H^p(\mathbb{D})$ with $\|f\|_p = \|h\|_p$. Since h has no zeros on $\mathbb{D}$, there is an analytic function ϕ such that $h = e^\phi$. In other words, $\log h$ is well-defined on $\mathbb{D}$. Put $g = e^{p\phi/2}$ and note that $\|g\|_2^2 = \|h\|_p^p = \|f\|_p^p$.

Theorem 1.10 is the first step toward an important factorization theorem in Hardy spaces. According to Theorem 1.10, we can extract the zeros of $f \in H^p(\mathbb{D})$, $0 < p \leq \infty$, as a Blaschke factor. The following result shows that there might still be a singular inner factor in f.

Theorem 1.12 (canonical factorization theorem) *Let $f \in H^p(\mathbb{D})$, $0 < p \leq \infty$, $f \not\equiv 0$. Then we have the unique factorization*

$$f = B S h,$$

where B is the Blaschke product formed with the zeros of f,

$$S(z) = \exp\left(-\int_{\mathbb{T}} \frac{e^{it} + z}{e^{it} - z} \, d\sigma(e^{it})\right), \qquad (z \in \mathbb{D}),$$

is a singular inner factor formed with a finite, positive and singular Borel measure σ on $\mathbb{T}$, and h is given by the formula

$$h(z) = \exp\left(\frac{1}{2\pi} \int_0^{2\pi} \frac{e^{it} + z}{e^{it} - z} \, \log|f(e^{it})| \, dt\right), \qquad (z \in \mathbb{D}).$$

Moreover, $h \in H^p(\mathbb{D})$ and

$$\|f\|_p = \|h\|_p.$$

Proof. The property (1.8) enables us to define h as above. The function h is analytic on $\mathbb{D}$ and satisfies the crucial identity

$$\log|h(re^{i\theta})| = \frac{1}{2\pi} \int_0^{2\pi} \frac{1 - r^2}{1 + r^2 - 2r\cos(\theta - t)} \, \log|f(e^{it})| \, dt. \qquad (1.21)$$

Therefore, by (1.9), we have

$$|f(z)| \leq |h(z)|, \qquad (z \in \mathbb{D}), \tag{1.22}$$

and, by (1.5),

$$\lim_{r \to 1} \log |h(re^{i\theta})| = \log |f(e^{i\theta})| \tag{1.23}$$

for almost all $e^{i\theta} \in \mathbb{T}$.

The identity (1.21) implies

$$|h(re^{i\theta})|^p \leq \frac{1}{2\pi} \int_0^{2\pi} \frac{1 - r^2}{1 + r^2 - 2r\cos(\theta - t)} \, |f(e^{it})|^p \, dt,$$

and thus $h \in H^p(\mathbb{D})$ with $\|h\|_p \leq \|f\|_p$, while (1.22) shows that $\|f\|_p \leq \|g\|_p$. Hence, the norm equality holds.

Put $\phi = f/h$. Since h has no zeros in $\mathbb{D}$, ϕ is an analytic function on the open unit disc. Moreover, by (1.22), $\phi \in H^\infty(\mathbb{D})$, and by (1.23),

$$\lim_{r \to 1} |\phi(re^{i\theta})| = 1$$

for almost all $e^{i\theta} \in \mathbb{T}$. In other words, ϕ is an inner function. According to Theorem 1.10, $\phi = BS$, where $S \in H^\infty(\mathbb{D})$ and is free of zeros in $\mathbb{D}$. For almost all $e^{i\theta} \in \mathbb{T}$ the radial limits of ϕ, B and S exist, and the radial limits of ϕ and B are unimodular. Hence S is also an inner function.

Since S is inner and free of zeros on $\mathbb{D}$, $-\log |S(z)|$ is a positive harmonic function on $\mathbb{D}$. Thus, according to Corollary 1.2, there is a positive Borel measure on $\mathbb{T}$ such that

$$\log |S(re^{i\theta})| = - \int_{\mathbb{T}} \frac{1 - r^2}{1 + r^2 - 2r\cos(\theta - t)} \, d\sigma(t). \tag{1.24}$$

On the one hand, since S is inner,

$$\lim_{r \to 1} \log |S(re^{i\theta})| = 0,$$

and on the other hand, by Theorem 1.3,

$$\lim_{r \to 1} \log |S(re^{i\theta})| = -2\pi \, \sigma'(e^{i\theta})$$

for almost all $e^{i\theta} \in \mathbb{T}$. Hence, σ is singular with respect to Lebesgue measure. Finally, the identity (1.24) can be rewritten as

$$\Re \log(S(z)) = \Re \left(- \int_{\mathbb{T}} \frac{e^{it} - z}{e^{it} + z} \, d\sigma(t) \right).$$

Therefore, there is a real constant β such that

$$S(z) = e^{i\beta} \, \exp \left(- \int_{\mathbb{T}} \frac{e^{it} - z}{e^{it} + z} \, d\sigma(t) \right).$$

The constant $e^{i\beta}$ can be absorbed into the Blaschke product.

In the canonical factorization described above, the function h is called the *outer part* of f and BS is called the *inner part* of f. Sometimes, instead of h and BS we will respectively write O_f and I_f. Hence, a function $f \in H^p(\mathbb{D})$, $0 < p \leq \infty$, is called outer if its inner part is a unimodular constant. Therefore, f is outer if and only if it has no zeros on $\mathbb{D}$ and the harmonic function $\log |f|$ is given by the Poisson integral of its boundary values, i.e.

$$\log |f(re^{i\theta})| = \frac{1}{2\pi} \int_0^{2\pi} \frac{1 - r^2}{1 + r^2 - 2r\cos(\theta - t)} \ \log |f(e^{it})| \ dt, \qquad (re^{i\theta} \in \mathbb{D}).$$

Note that if f, g are in some Hardy spaces (not necessarily the same) and $g \not\equiv 0$, then O_f/O_g is also an outer function and

$$\frac{O_f}{O_g} = O_{f/g}. \tag{1.25}$$

Up to now, we have seen two classes of inner functions: Blaschke products and singular inner functions. Clearly any product of the form BS is also an inner function. But more importantly, the canonical factorization theorem shows that any inner function is of this form. This fact is implicitly stated in Theorem 1.12. But, because of its importance for us, we repeat it again below as a corollary.

Corollary 1.13 *Let ϕ be an inner function for the open unit disc $\mathbb{D}$. Let B the Blaschke product formed with the zeros of ϕ. Then there is a finite, positive, and singular Borel measure σ on $\mathbb{T}$ such that*

$$\phi = B \, S_\sigma.$$

1.5 A Characterization of Blaschke Products

In Corollary 1.13, we saw that an inner function ϕ decomposes as $\phi = BS_\sigma$, where B is a Blaschke product and S_σ is a singular inner function. The factor B reduces to a unimodular constant if and only if ϕ has no zeros on $\mathbb{D}$. Naturally, we may propose a similar question for the singular factor S_σ. In other words, under which extra conditions, an inner function has to be a Blaschke product? In this section we study two conditions of this type.

Theorem 1.14 *Let ϕ be an inner function for the open unit disc $\mathbb{D}$. Then the limit*

$$\lim_{r \to 1} \int_0^{2\pi} \log |\phi(re^{i\theta})| \ d\theta$$

exists. Moreover, ϕ is a Blaschke product if and only if

$$\lim_{r \to 1} \int_0^{2\pi} \log \left| \phi(re^{i\theta}) \right| \, d\theta = 0.$$

Proof. By Corollary 1.13, we have the decomposition $\phi = BS_\sigma$. Hence, by (1.20),

$$\log \left| \phi(re^{i\theta}) \right| = \log \left| B(re^{i\theta}) \right| - \frac{1}{2\pi} \int_{\mathbb{T}} \frac{1 - r^2}{1 + r^2 - 2r \cos(\theta - t)} \, d\sigma(t),$$

which, by (1.1), implies

$$\int_0^{2\pi} \log \left| \phi(re^{i\theta}) \right| \, d\theta = \int_0^{2\pi} \log \left| B(re^{i\theta}) \right| \, d\theta - \int_{\mathbb{T}} d\sigma(t).$$

Therefore, by Lemma 1.6, the required limit exists and is equal to

$$\lim_{r \to 1} \int_0^{2\pi} \log \left| \phi(re^{i\theta}) \right| \, d\theta = \lim_{r \to 1^-} \int_0^{2\pi} \log \left| B(re^{i\theta}) \right| \, d\theta - \int_{\mathbb{T}} d\sigma(t) = -\sigma(\mathbb{T}).$$

The above identities show that

$$\lim_{r \to 1} \int_0^{2\pi} \log \left| \phi(re^{i\theta}) \right| \, d\theta = 0,$$

holds if and only if $\sigma(\mathbb{T}) = 0$. But, since σ is positive, $\sigma(\mathbb{T}) = 0$ is equivalent to $\sigma \equiv 0$.

In the preceding result, we considered the integral means of ϕ to detect the Blaschke products. In the following result, we consider the radial limits.

Theorem 1.15 *Let ϕ be an inner function for the unit disc $\mathbb{D}$. Suppose that*

$$\lim_{r \to 1} \left| \phi(re^{i\theta}) \right| \neq 0$$

for all $e^{i\theta} \in \mathbb{T}$. Then ϕ is a Blaschke product.

Remark. The assumption

$$\lim_{r \to 1} \left| \phi(re^{i\theta}) \right| \neq 0$$

means that either the radial limit $\lim_{r \to 1} \left| \phi(re^{i\theta}) \right|$ does not exist, or it exists but its value is not equal to zero.

Proof. By Corollary 1.13, we have the decomposition $\phi = BS_\sigma$, and thus

$$\left| \phi(re^{i\theta}) \right| \leq \left| S_\sigma(re^{i\theta}) \right|, \qquad (re^{i\theta} \in \mathbb{D}).$$

Suppose that $\sigma \neq 0$. Then, by Lemma 1.8, there is at least one point $e^{i\theta} \in \mathbb{T}$ such that

$$\lim_{r \to 1} S_\sigma(re^{i\theta}) = 0.$$

Thus, at this point, we would have

$$\lim_{r \to 1} |\phi(re^{i\theta})| = 0$$

which is a contradiction.

Example 1.16   Theorem 1.15 enables us to detect some nontrivial Blaschke products [7]. Let σ be the Dirac measure with unit mass at the point one. Then the formula (1.19) gives the atomic inner function

$$S(z) = \exp\left(-\frac{1+z}{1-z} \right).$$

It is straightforward to see that

$$\lim_{r \to 1^-} S(re^{i\theta}) = S(e^{i\theta}) = \exp\left(-i \cot(\theta/2) \right)$$

for every $e^{i\theta} \in \mathbb{T} \setminus \{1\}$, and that

$$\lim_{r \to 1^-} S(r) = 0.$$

Fix any $\alpha \in \mathbb{D}$, $\alpha \neq 0$, and let

$$\phi(z) = \frac{\alpha - S(z)}{1 - \bar{\alpha}\, S(z)}, \qquad (z \in \mathbb{D}).$$

Appealing to the elementary properties of the Blaschke factor $(\alpha - w)/(1 - \bar{\alpha}\, w)$, we see that

$$|\phi(z)| < 1, \qquad (z \in \mathbb{D}),$$

and

$$\lim_{r \to 1^-} \phi(re^{i\theta}) = \frac{\alpha - S(e^{i\theta})}{1 - \bar{\alpha}\, S(e^{i\theta})} \in \mathbb{T}$$

for every $e^{i\theta} \neq 1$, and that

$$\lim_{r \to 1} \phi(r) = \alpha \neq 0.$$

Hence, ϕ is an inner function. Moreover, by Theorem 1.14, ϕ is in fact a Blaschke product. The zeros of ϕ are the solutions of the equation $S(z) = \alpha$, i.e.

$$\exp\left(-\frac{1+z}{1-z} \right) = \alpha.$$

By an elementary calculation, the solution of the above equation are

$$z_n = \frac{\log|\alpha| + i(\arg\alpha + 2n\pi) + 1}{\log|\alpha| + i(\arg\alpha + 2n\pi) - 1}, \qquad (n \in \mathbb{Z}).$$

In particular, for $\alpha = 1/e$, the solutions are

$$z_n = \frac{in\pi}{in\pi - 1}, \qquad (n \in \mathbb{Z}).$$

Note that, without appealing to Theorem 1.15, it is not straightforward that

$$\phi(z) = \frac{\alpha - e^{-\frac{1+z}{1-z}}}{1 - \bar{\alpha}\, e^{-\frac{1+z}{1-z}}}, \qquad (\alpha \in \mathbb{D} \setminus \{0\}),$$

is a Blaschke product.

1.6 The Nevanlinna Class $\mathcal{N}$ and Its Subclass $\mathcal{N}^+$

The *Nevanlinna* class $\mathcal{N}$ is the family of all analytic functions f on the open unit disc which satisfy the growth restriction

$$\sup_{0 \leq r < 1} \int_{-\pi}^{\pi} \log^+ |f(re^{i\theta})|\, d\theta < \infty. \tag{1.26}$$

It is well-known that $f \in \mathcal{N}$ if and only if f is the quotient of two bounded functions. In the light of the canonical factorization theorem, this characterization immediately implies the following properties of an element $f \in \mathcal{N}$.

(i) The zeros of f satisfy the Blaschke condition. This property also follows from Lemma 1.9.

(ii) For almost all $e^{i\theta} \in \mathbb{T}$, the radial limits

$$f(e^{i\theta}) = \lim_{r \to 1} f(re^{i\theta})$$

exist and are finite.

(iii) If $f \not\equiv 0$, then $\log|f| \in L^1(\mathbb{T})$.

(iv) There is a finite signed singular Borel measure σ on $\mathbb{T}$ such that

$$f = B\, S_\sigma\, O_{|f|},$$

where B is the Blaschke product formed with the zeros of f.

On the other hand, if B is any Blaschke product, σ is any finite signed singular Borel measure on $\mathbb{T}$, h is any positive measurable function with $\log h \in L^1(\mathbb{T})$, then

$$B \, S_\sigma \, O_h \in \mathcal{N}.$$

We emphasize that the measure σ appearing in the canonical decomposition of functions in the Nevanlinna class is not necessarily positive. Hence, we define the proper subclass

$$\mathcal{N}^+ = \{\, f \in \mathcal{N} : f = B \, S_\sigma \, O_{|f|} \text{ with } \sigma \geq 0 \,\}.$$

Based on the canonical factorization theorem, we have

$$\bigcup_{p>0} H^p(\mathbb{D}) \subset \mathcal{N}^+,$$

and the inclusion is proper. The main advantage of $\mathcal{N}^+$ over $\mathcal{N}$ is the following result.

Theorem 1.17 *Let $f \in \mathcal{N}^+$. Then $f \in H^p(\mathbb{D})$, $0 < p \leq \infty$, if and only if $f \in L^p(\mathbb{T})$. In particular, if $f \in H^s(\mathbb{D})$ for some $s \in (0, \infty]$ and $f \in L^p(\mathbb{T})$, for some $0 < p \leq \infty$, then $f \in H^p(\mathbb{D})$.*

Proof. Since $f \in \mathcal{N}^+$, it has the canonical factorization $f = B \, S_\sigma \, O_{|f|}$, where B is the Blaschke product formed with the zeros of f and σ is a finite positive singular Borel measure on $\mathbb{T}$. Therefore, $\phi = B \, S_\sigma$ is an inner function. Hence, it is enough to show that $O_{|f|} \in H^p(\mathbb{D})$. But, $O_{|f|} \in H^p(\mathbb{D})$ if and only if $f \in L^p(\mathbb{T})$.

The function

$$f(z) = \exp\left\{ \frac{1+z}{1-z} \right\}, \qquad (z \in \mathbb{D}),$$

which was introduced at the end of Sect. 1.2 is such that $f \in \mathcal{N}$ with the unimodular boundary values

$$f(e^{i\theta}) = e^{i \cot(\theta/2)}, \qquad (e^{i\theta} \in \mathbb{T}).$$

However, $f \notin H^\infty(\mathbb{D})$. This example shows that in Theorem 1.17, the assumption $f \in \mathcal{N}^+$ cannot be weakened by replacing it with $f \in \mathcal{N}$.

The following result provides a characterization of the elements of $\mathcal{N}^+$ in the Nevanlinna class $\mathcal{N}$.

Theorem 1.18 *Let $f \in \mathcal{N}$. Then $f \in \mathcal{N}^+$ if and only if*

$$\lim_{r \to 1} \int_{-\pi}^{\pi} \log^+ |f(re^{i\theta})| \, d\theta = \int_{-\pi}^{\pi} \log^+ |f(e^{i\theta})| \, d\theta. \qquad (1.27)$$

Proof. Suppose that $f \in \mathcal{N}^+$. Hence $f = B \, S_\sigma \, O_{|f|}$, where σ is a finite positive singular Borel measure on $\mathbb{T}$. Therefore, $|f| \leq |O_{|f|}|$, which implies

$$\log^+ |f(re^{i\theta})| \leq \frac{1}{2\pi} \int_{-\pi}^{\pi} \frac{1-r^2}{1+r^2 - 2r\cos(\theta - t)} \log^+ |f(e^{it})| \, dt.$$

Thus, by Fubini's theorem and (1.1),

$$\int_{-\pi}^{\pi} \log^+ |f(re^{i\theta})| \, d\theta \leq \int_{-\pi}^{\pi} \log^+ |f(e^{it})| \, dt, \qquad (0 \leq r < 1).$$

The quantity on the left side is an increasing function of r, and by Fatou's lemma,

$$\int_{-\pi}^{\pi} \log^+ |f(e^{i\theta})| \, d\theta \leq \liminf_{r \to 1} \int_{-\pi}^{\pi} \log^+ |f(re^{i\theta})| \, d\theta.$$

The preceding two inequalities show that (1.27) holds.

The other direction is a bit more delicate. Suppose that (1.27) holds. Let $f = B \, S_\sigma \, O_{|f|}$ be the canonical decomposition of $f \in \mathcal{N}$. Write $g = S_\sigma \, O_{|f|}$. Since $f = Bg$, it is enough to prove that $g \in \mathcal{N}^+$. We have $|f| = |g|$ and

$$\log |B(z)| + \log^+ |g(z)| \leq \log^+ |f(z)| \leq \log^+ |g(z)|.$$

Thus, by Theorem 1.14 and our assumption (1.27),

$$\lim_{r \to 1} \int_{-\pi}^{\pi} \log^+ |g(re^{i\theta})| \, d\theta = \int_{-\pi}^{\pi} \log^+ |g(e^{i\theta})| \, d\theta. \qquad (1.28)$$

The function g has the integral representation

$$\log |g(re^{i\theta})| = \frac{1}{2\pi} \int_{\mathbb{T}} \frac{1-r^2}{1+r^2 - 2r\cos(\theta - t)} \, d\lambda(e^{it}), \qquad (re^{i\theta} \in \mathbb{D}),$$

where

$$d\lambda(e^{it}) = \log |g(e^{it})| \, dt - d\sigma(e^{it}).$$

Hence, the measures $\log |G(re^{it})| \, dt$ converge in the weak* topology to $d\lambda(e^{it})$. Take any sequence $r_n > 0$, $n \geq 1$, such that $r_n \to 1$. Since the sequences $(\log^+ |G(r_n e^{it})| \, dt)_{n \geq 1}$ and $(\log^- |G(r_n e^{it})| \, dt)_{n \geq 1}$ are uniformly bounded in the Banach space $\mathcal{M}(\mathbb{T})$, there is a subsequence $(n_k)_{k \geq 1}$ and two positive measures $\lambda_1, \lambda_2 \in \mathcal{M}(\mathbb{T})$ such that

$$\log^+ |G(r_{n_k} e^{it})| \, dt \longrightarrow d\lambda_1(e^{it}) \quad \text{and} \quad \log^- |G(r_{n_k} e^{it})| \, dt \longrightarrow d\lambda_2(e^{it})$$

in the weak-star topology. Hence, the measure λ is given by $\lambda = \lambda_1 - \lambda_2$. Our main task is to show that λ_1 is absolutely continuous with respect to the Lebesgue measure. This fact is equivalent to saying that $\sigma \geq 0$, and thus the theorem would be proved.

Let E be any Borel subset of $\mathbb{T}$. Then, by Fatou's lemma,

$$\int_E \log^+ |g(e^{it})| \, dt \leq \liminf_{r \to 1} \int_E \log^+ |g(re^{it})| \, dt$$

and

$$\int_{\mathbb{T} \backslash E} \log^+ |g(e^{it})| \, dt \leq \liminf_{r \to 1} \int_{\mathbb{T} \backslash E} \log^+ |g(re^{it})| \, dt.$$

If, in any one of the last two inequalities, the strict inequality holds, then we add them up and obtain a strict inequality, with integrals on $\mathbb{T}$, which contradicts (1.28). Hence,

$$\liminf_{r \to 1} \int_E \log^+ |g(re^{it})| \, dt = \int_E \log^+ |g(e^{it})| \, dt$$

for all Borel subsets E of $\mathbb{T}$. In particular, if we take $r = r_{n_k}$, $k \longrightarrow \infty$, then we see that the measures $\log^+ |g(r_{n_k} e^{it})| \, dt$ converge to $\log^+ |g(e^{it})| \, dt$ in the weak-star topology. Therefore,

$$d\lambda_1(e^{it}) = \log^+ |g(e^{it})| \, dt \geq 0.$$

1.7 Bergman Spaces

Fix $0 < p < 1$. The Banach space $B^p(\mathbb{D})$ consists of all analytic functions on $\mathbb{D}$ for which the quantity

$$\|f\|_{B^p} = \int_0^1 \int_0^{2\pi} |f(re^{i\theta})| \, (1-r)^{\frac{1}{p}-2} \, dr d\theta$$

is finite. Some authors prefer to replace $dr d\theta$ by $r dr d\theta$, or to multiply the norm by a normalization constant. This comment also applies to other norms which are defined below. It is easy to observe that

$$0 < p < q < 1 \Longrightarrow B^q(\mathbb{D}) \subset B^p(\mathbb{D}).$$

It is also well-known that

$$B^p(\mathbb{D}) \subset H^p(\mathbb{D}), \qquad (0 < p < 1).$$

But, we do not need this result in the following.

The classical *Bergman space* $A^p(\mathbb{D})$, $0 < p < \infty$, consists of all analytic functions f on $\mathbb{D}$ such that the quantity

$$\|f\|_{A^p} = \left(\frac{1}{\pi} \int_0^1 \int_0^{2\pi} |f(re^{i\theta})|^p \, r dr d\theta \right)^{\frac{1}{p}}$$

is finite. In the definition above, the constant is adjusted such that the norm of the constant function $f \equiv 1$ is precisely 1. Note that

$$A^1(\mathbb{D}) = B^{\frac{1}{2}}(\mathbb{D}).$$

The weighted Bergman spaces A_γ^p are defined via the growth restriction

$$\|f\|_{A_\gamma^p} = \left(\frac{1+\gamma}{\pi} \int_0^1 \int_0^{2\pi} |f(re^{i\theta})|^p \, r(1-r^2)^\gamma dr \, d\theta \right)^{\frac{1}{p}} < \infty,$$

where $\gamma > -1$. Hence, we have

$$A^p = A_0^p.$$

It is easy to observe that

$$0 < p < q < \infty \Longrightarrow A^q(\mathbb{D}) \subset A^p(\mathbb{D}),$$

or even more generally, for a fixed γ,

$$0 < p < q < \infty \Longrightarrow A_\gamma^q(\mathbb{D}) \subset A_\gamma^p(\mathbb{D}).$$

If $0 < p < 1$, then

$$d(f,g) = \|f - g\|_{A_\gamma^p}^p$$

defines a complete translation-invariant metric on $A_\gamma^p(\mathbb{D})$, and thus $A_\gamma^p(\mathbb{D})$ is a Fréchet space in this case. But, for $1 \leq p < \infty$, $\| \cdot \|_{A_\gamma^p}$ is a well-defined complete norm on $A_\gamma^p(\mathbb{D})$ and thus $A_\gamma^p(\mathbb{D})$ is a Banach space. In particular, for $p = 2$, the norm is produced by the inner product

$$\langle f,g \rangle_{A_\gamma^2} = \frac{1}{\pi(1+\gamma)} \int_0^1 \int_0^{2\pi} f(re^{i\theta}) \, \overline{g(re^{i\theta})} \, r(1-r^2)^\gamma dr \, d\theta.$$

In other words, $A_\gamma^2(\mathbb{D})$ is a Hilbert space. If f and g have the Taylor series representations

$$f(z) = \sum_{n=0}^\infty a_n z^n \qquad \text{and} \qquad g(z) = \sum_{n=0}^\infty b_n z^n$$

on $\mathbb{D}$, then either by direct verification or by applying the Parseval identity, we obtain

$$\langle f,g \rangle_{A_\gamma^2} = \sum_{n=0}^\infty \frac{n! \, \Gamma(\gamma+2)}{\Gamma(\gamma+n+2)} \, a_n \bar{b}_n. \tag{1.29}$$

Hence,

$$\|f\|_{A_\gamma^2}^2 = \sum_{n=0}^\infty \frac{n! \, \Gamma(\gamma+2)}{\Gamma(\gamma+n+2)} \, |a_n|^2. \tag{1.30}$$

In particular, in the classical case $A^2(\mathbb{D})$, we have

$$\langle f, g \rangle_{A^2} = \sum_{n=0}^{\infty} \frac{a_n \bar{b}_n}{n+1}, \tag{1.31}$$

and

$$\|f\|_{A^2}^2 = \sum_{n=0}^{\infty} \frac{|a_n|^2}{n+1}. \tag{1.32}$$

Theorem 1.19 (Hardy–Littlewood [25]) *Let $0 < p < \infty$. Then every function $f \in H^p(\mathbb{D})$ also belongs to $A^{2p}(\mathbb{D})$ and satisfies $\|f\|_{A^{2p}} \le \|f\|_{H^p}$, with equality if and only if*

$$f(z) = \frac{C}{(1 - \lambda z)^{2/p}},$$

where $\lambda \in \mathbb{D}$ and $C \in \mathbb{C}$ are arbitrary constants.

Proof (Vukotić [45]). We first consider the case $p = 2$. If f has the Taylor series representation

$$f(z) = \sum_{n=0}^{\infty} a_n z^n$$

then f^2 has the Taylor series

$$f^2(z) = \sum_{n=0}^{\infty} \left(\sum_{k=0}^{n} a_k a_{n-k} \right) z^n.$$

Therefore, by (1.32),

$$\|f\|_{A^4}^4 = \|f^2\|_{A^2}^2 = \sum_{n=0}^{\infty} \frac{1}{n+1} \left| \sum_{k=0}^{n} a_k a_{n-k} \right|^2.$$

The Cauchy-Schwarz inequality gives

$$\left| \sum_{k=0}^{n} a_k a_{n-k} \right|^2 \le (n+1) \sum_{k=0}^{n} |a_k|^2 |a_{n-k}|^2. \tag{1.33}$$

Hence,

$$\|f\|_{A^4}^4 \le \sum_{n=0}^{\infty} \sum_{k=0}^{n} |a_k|^2 |a_{n-k}|^2 = \left(\sum_{m=0}^{\infty} |a_m|^2 \right)^2 = \|f\|_{H^2}^4.$$

The equality holds if and only if we have equality in (1.33) for all $n \ge 0$, and the latter happens if and only if

$$a_k \, a_{n-k} = c_n, \qquad (0 \leq k \leq n),$$

where c_n does not depend on k. In particular, we must have $a_n^2 = a_0 \, a_{2n} = c_{2n}$. This observation shows that $a_0 = 0$ forces all other coefficients to be zero and thus $f = 0$. If $a_0 \neq 0$, we use $a_n \, a_0 = a_{n-1} \, a_1 = c_n$ to deduce

$$a_n = \lambda \, a_{n-1}, \qquad (n \geq 1),$$

where $\lambda = a_1/a_0$. Hence,

$$a_n = \lambda^n \, a_0, \qquad (n \geq 0).$$

With this choice of coefficients, we obtain

$$f(z) = \frac{a_0}{1 - \lambda z}, \qquad (1.34)$$

and, for this function, equality holds in (1.33).

For other values of p, we use the Riesz technique (Corollary 1.11). Write $f = B \, g^{2/p}$, with $g \in H^2(\mathbb{D})$ and $\|f\|_p^p = \|g\|_2^2$. Then, by the above discussion,

$$\|f\|_{A^{2p}}^{2p} \leq \|g\|_{A^4}^4 \leq \|g\|_{H^2}^4 = \|f\|_{H^p}^{2p}.$$

The first inequality becomes an equality if and only if there is no Blaschke factor. The second is an equality if and only if g has the form (1.34). Hence, $\|f\|_{A^{2p}} = \|f\|_{H^p}$ if and only if

$$f(z) = \frac{C}{(1 - \lambda z)^{2/p}}.$$

The above proof shows that, when $1 \leq p < \infty$, the injection mapping

$$H^p(\mathbb{D}) \longrightarrow A^{2p}(\mathbb{D})$$
$$f \longmapsto f$$

has norm one. Moreover, this result is sharp in the sense that the exponent $2p$ is the best possible choice. More precisely,

$$H^p(\mathbb{D}) \not\subset A^q(\mathbb{D}),$$

for any $q > 2p$. This is shown in the Example 4.5. For more information on Bergman spaces see [18] and [26].

Chapter 2
The Exceptional Set of an Inner Function

2.1 Frostman Shifts and the Exceptional Set $\mathcal{E}_\varphi$

For a fixed $w \in \mathbb{D}$, the mapping

$$\tau_w(z) = \frac{w - z}{1 - \bar{w}\, z}, \qquad (z \in \mathbb{D}),$$

is an automorphism of the open unit disc $\mathbb{D}$. Hence, for an inner function φ,

$$\varphi_w(z) = \frac{w - \varphi(z)}{1 - \bar{w}\, \varphi(z)}, \qquad (z \in \mathbb{D}),$$

maps $\mathbb{D}$ into itself. Hence, at least φ_w is an element of the closed unit ball of H^∞. Moreover, for almost all $e^{i\theta} \in \mathbb{T}$,

$$\lim_{r \to 1} \varphi_w(re^{i\theta}) = \frac{w - \varphi(e^{i\theta})}{1 - \bar{w}\, \varphi(e^{i\theta})} = -\overline{\varphi(e^{i\theta})}\, \frac{w - \varphi(e^{i\theta})}{\bar{w} - \overline{\varphi(e^{i\theta})}} \in \mathbb{T}.$$

Therefore, for each $w \in \mathbb{D}$, φ_w is in fact an inner function. What is not obvious is that φ_w has a *good chance* to be a Blaschke product. In other words, the *exceptional set*

$$\mathcal{E}_\varphi = \{\, w \in \mathbb{D} : \varphi_w \text{ is not a Blaschke product} \,\}$$

is *small*. What does small mean? In this section, we show that the Lebesgue measure of $\mathcal{E}_\varphi$ is zero. In Sect. 2.4, we refine this result by showing that the logarithmic capacity of $\mathcal{E}_\varphi$ is zero.

In view of the following result, the functions φ_w, $w \in \mathbb{D}$, are called the *Frostman shifts* of φ. The following version of Frostman's theorem says that the two-dimensional Lebesgue measure of $\mathcal{E}_\varphi$ is zero.

Theorem 2.1 (Frostman [23]) *Let φ be an inner function for the open unit disc $\mathbb{D}$. Fix $0 < \rho < 1$, and define*

J. Mashreghi, *Derivatives of Inner Functions*, Fields Institute Monographs 31,
DOI 10.1007/978-1-4614-5611-7_2, © Springer Science+Business Media New York 2013

$$\mathcal{E}_\rho(\varphi) = \{\, e^{i\vartheta} \in \mathbb{T} : \varphi_{\rho e^{i\vartheta}} \text{ is not a Blaschke product}\,\}.$$

Then $\mathcal{E}_\rho(\varphi)$ has (one-dimensional) Lebesgue measure zero.

Proof. For each $\alpha \in \mathbb{D}$, we have

$$\frac{1}{2\pi} \int_0^{2\pi} \log \left| \frac{\rho e^{i\vartheta} - \alpha}{1 - \rho e^{-i\vartheta}\alpha} \right| d\vartheta = \max(\log \rho, \log |\alpha|). \tag{2.1}$$

Since φ is inner, we can replace α by $\varphi(re^{it})$ and then integrate with respect to t. Hence,

$$\frac{1}{2\pi} \int_0^{2\pi} \left(\int_0^{2\pi} \log \left| \varphi_{\rho e^{i\vartheta}}(re^{it}) \right| d\vartheta \right) dt = \int_0^{2\pi} \max(\log \rho, \log |\varphi(re^{it})|)\, dt,$$

where

$$\varphi_{\rho e^{i\vartheta}}(z) = \frac{\rho e^{i\vartheta} - \varphi(z)}{1 - \rho e^{-i\vartheta}\,\varphi(z)}, \qquad (z \in \mathbb{D}).$$

Since ρ is fixed and $|\varphi| \le 1$, the family

$$f_r(e^{it}) = \max(\log \rho, \log |\varphi(re^{it})|), \qquad (e^{it} \in \mathbb{T}),$$

where the parameter r runs through $[0, 1)$, is uniformly bounded. More explicitly, we have

$$\log \rho \le f_r(e^{it}) \le 0, \qquad (e^{it} \in \mathbb{T}).$$

Moreover,

$$\lim_{r \to 1} f_r(e^{it}) = \max(\log \rho, \lim_{r \to 1} \log |\varphi(re^{it})|) = \max(\log \rho, 0) = 0$$

for almost all $e^{it} \in \mathbb{T}$. Hence, by the dominated convergence theorem,

$$\lim_{r \to 1} \int_0^{2\pi} f_r(e^{it})\, dt = 0.$$

Therefore,

$$\lim_{r \to 1} \int_0^{2\pi} \left(\int_0^{2\pi} \log \left| \varphi_{\rho e^{i\vartheta}}(re^{it}) \right| d\vartheta \right) dt = 0.$$

But, considering the fact that the integrand $\log |\varphi_{\rho e^{i\vartheta}}|$ is negative, in the light of Fubini's theorem, we can change the order of integration and write

$$\lim_{r \to 1} \int_0^{2\pi} \left(\int_0^{2\pi} \log \left| \varphi_{\rho e^{i\vartheta}}(re^{it}) \right| dt \right) d\vartheta = 0. \tag{2.2}$$

Now, it is the time to appeal to Fatou's lemma. For simplicity of notation, put

$$M(r, w) = \int_0^{2\pi} -\log \left| \varphi_w(re^{it}) \right| \, dt.$$

According to Theorem 1.14, we know that, for each fixed point $w \in \mathbb{D}$, the limit

$$\lim_{r \to 1} M(r, w)$$

exists and is a positive number. Moreover, by Fatou's lemma,

$$\int_0^{2\pi} \left(\liminf_{r \to 1} M(r, \rho e^{i\vartheta}) \right) d\vartheta \le \liminf_{r \to 1} \int_0^{2\pi} M(r, \rho e^{i\vartheta}) \, d\vartheta.$$

Hence, by (2.2) and that $M \ge 0$,

$$\int_0^{2\pi} \left(\lim_{r \to 1} M(r, \rho e^{i\vartheta}) \right) d\vartheta = 0,$$

which implies

$$\lim_{r \to 1} M(r, \rho e^{i\vartheta}) = 0$$

for almost all $\vartheta \in [0, 2\pi]$. Therefore, again by Theorem 1.14, $\varphi_{\rho e^{i\vartheta}}$ is a Blaschke product for almost all $\vartheta \in [0, 2\pi]$. This means that $\mathcal{E}_\rho(\varphi)$ has Lebesgue measure zero.

The preceding result immediately implies an interesting approximation theorem. It shows that the set of Blaschke products is uniformly dense in the set of all inner functions.

Corollary 2.2 (Frostman [23]) *Let φ be an inner function for the open unit disc $\mathbb{D}$. Then, given $\varepsilon > 0$, there is a Blaschke product B such that*

$$\|\varphi - B\|_{H^\infty} < \varepsilon.$$

Proof. Take $\rho \in (0, 1)$ small enough such that $2\rho/(1 - \rho) < \varepsilon$. According to Theorem 2.1, on the circle $\{|z| = \rho\}$, there are many candidates $\rho e^{i\vartheta}$ such that $\varphi_{\rho e^{i\vartheta}}$ is a Blaschke product. Pick any one you wish. Then we have

$$|\varphi(z) + \varphi_{\rho e^{i\vartheta}}(z)| = \left| \frac{\rho e^{i\vartheta} - \rho e^{-i\vartheta}\varphi^2(z)}{1 - \rho e^{-i\vartheta}\varphi(z)} \right| \le \frac{2\rho}{1 - \rho} < \varepsilon$$

for all $z \in \mathbb{D}$. This simply means that

$$\|\varphi + \varphi_{\rho e^{i\vartheta}}\|_\infty < \varepsilon.$$

Hence, the Blaschke product $B = -\varphi_{\rho e^{i\vartheta}}$ fulfils the required inequality $\|\varphi - B\|_{H^\infty} < \varepsilon$.

2.2 Capacity

Let

$$\Phi_\alpha(r) = \begin{cases} 1/r^\alpha & \text{if } \alpha > 0, \\ \log 1/r & \text{if } \alpha = 0. \end{cases}$$

Let μ be a positive Borel measure with compact support in the complex plane $\mathbb{C}$. Then the *potential* function created by μ and the kernel Φ_α is defined by

$$P_{\alpha,\mu}(z) = \int_\mathbb{C} \Phi_\alpha(|z - w|) \, d\mu(w). \tag{2.3}$$

In particular, $P_\mu = P_{0,\mu}$ is the famous *logarithmic potential*. Since the kernel Φ_0 assumes both positive and negative values, the treatment of the logarithmic potential is slightly different from the others. It is easy to see that $P_{\alpha,\mu}$ is a superharmonic function with values in $(-\infty, \infty]$ on the complex plane $\mathbb{C}$ with

$$\Phi_\alpha(r) = \begin{cases} \|\mu\| \, (1 + o(1))/|z|^\alpha & \text{if } \alpha > 0, \\ \|\mu\| \, \log(1/|z|) \, (1 + o(1)) & \text{if } \alpha = 0, \end{cases}$$

as $z \longrightarrow \infty$.

In some applications, in particular when we treat subsets of the open unit disc $\mathbb{D}$, it is easier to work with the Green potential

$$G_{\alpha,\mu}(z) = \int_\mathbb{C} \Phi_\alpha\left(\left|\frac{w - z}{1 - \bar{w} z}\right|\right) \, d\mu(w). \tag{2.4}$$

The main advantage of this potential is that, as w ranges on a compact subset of $\mathbb{D}$, the quantity $|w - z|/|1 - \bar{w} z|$ uniformly tends to 1 as $|z| \longrightarrow 1$. Hence, in this situation, $G_{0,\mu}(z) \longrightarrow 0$. In other words, $G_{0,\mu}$ is continuous on an annulus around $\mathbb{T}$ and, moreover, $G_{0,\mu} \equiv 0$ on $\mathbb{T}$.

The quantity

$$\mathcal{E}_{\alpha,\mu} = \int_\mathbb{C} P_{\alpha,\mu}(z) \, d\mu(z) = \int_\mathbb{C} \int_\mathbb{C} \Phi_\alpha(|z - w|) \, d\mu(w) d\mu(z) \tag{2.5}$$

is called the *energy* of μ. Note that in (2.3) and (2.5), we can replace $\int_\mathbb{C}$ by $\int_{\text{supp } \mu}$ where supp μ is the support of μ. For a given compact set K, the quantities

$$\inf_\mu \left(\sup_{z \in K} P_{\alpha,\mu}(z) \right) \quad \text{and} \quad \inf_\mu \mathcal{E}_{\alpha,\mu}, \tag{2.6}$$

where in both cases the infimum is taken over all probability measures μ whose supports are in K, are important in our discussion. The *maximum principle*, one of the most essential results of potential theory, says that

$$\sup_{z \in K} P_{\alpha,\mu}(z) = \sup_{z \in \mathbb{C}} P_{\alpha,\mu}(z),$$

and thus, in (2.6) we can replace K by $\mathbb{C}$. Then, according to a celebrated theorem of Frostman in potential theory [22], the quantities introduced in (2.6) are equal, and hence we define

$$\Delta_\alpha(K) = \inf_\mu \left(\sup_{z \in K} P_{\alpha,\mu}(z) \right) = \inf_\mu \mathcal{E}_{\alpha,\mu}.$$

Using this fact, the C_α-*capacity* of K, denoted by $C_\alpha(K)$, is defined by the equation

$$\Phi_\alpha\big(C_\alpha(K)\big) = \Delta_\alpha(K).$$

Since Φ_α is one-to-one, $C_\alpha(K)$ is well-defined. In other words, we have

$$C_\alpha(K) = \begin{cases} \Delta_\alpha(K)^{-\frac{1}{\alpha}} & \text{if } \alpha > 0, \\ e^{-\Delta_\alpha(K)} & \text{if } \alpha = 0. \end{cases} \tag{2.7}$$

The C_0-capacity is also referred to as the *logarithmic capacity*. It is worthwhile to mention that Frostman's theorem goes further and says that there is a particular measure μ, which is called the *equilibrium measure* and whose choice surely depends on K and α, such that

$$\Delta_\alpha(K) = \sup_{z \in K} P_{\alpha,\mu}(z) = \mathcal{E}_{\alpha,\mu}.$$

It is usually difficult to exactly evaluate $C_\alpha(K)$. As a matter of fact, in most applications, we simply want to know either $C_\alpha(K) = 0$ or $C_\alpha(K) > 0$. The following result gives a characterization of the case $C_\alpha(K) > 0$.

Theorem 2.3 *Let K be a compact subset of the open unit disc $\mathbb{D}$, and let $\alpha \geq 0$. Then the following assertions are equivalent.*

 (i) *$C_\alpha(K) > 0$.*
 (ii) *There is a positive Borel measure μ, with $\mu \neq 0$, whose support is in K and the potential $P_{\alpha,\mu}$ is bounded above on $\mathbb{C}$.*
 (iii) *There is a positive Borel measure μ, with $\mu \neq 0$, whose support is in K and the Green potential $G_{\alpha,\mu}$ is bounded above on $\mathbb{C}$.*

Proof. $(i) \Longleftrightarrow (ii)$: This is a direct consequence of (2.7).
   $(ii) \Longleftrightarrow (iii)$: It follows from the estimations

$$0 < 1 - d \leq |1 - \bar{w}\,z| < 2,$$

where $d = \max_{z \in K} |z|$.

An arbitrary set $E \subset \mathbb{C}$ is said to have positive α-capacity if there is a compact subset of E whose α-capacity is positive. The following result shows that a Borel set of logarithmic capacity zero has the (two-dimensional) Lebesgue measure zero. Knowing that the logarithmic capacity of an interval is positive, we see that the family of Borel subsets of $\mathbb{C}$ with logarithmic capacity zero is a *proper* subclass of the family of all Borel subsets of Lebesgue measure zero.

Corollary 2.4 *Let E be a Borel subset of $\mathbb{C}$ whose two-dimensional Lebesgue measure is positive. Then, for each $0 \leq \alpha < 2$, we have $C_\alpha(E) > 0$.*

Proof. The result essentially follows from the fact that the functions $\log 1/|z|$ and $1/|z|^\alpha$, for $0 < \alpha < 2$, are locally integrable.

By regularity, there is a compact subset K in E whose two-dimensional Lebesgue measure is positive. Without loss of generality, we can assume that $d = \mathrm{diam}(K) < 1$. Denoting the Lebesgue measure by m, define μ by

$$\mu(A) = m(A \cap K),$$

where A is a Borel subset of $\mathbb{C}$. Then μ is a positive Borel measure, with $\mu \neq 0$ and $\mathrm{Supp}\, \mu = K$, and

$$P_{\alpha,\mu}(z) = \int_{\mathbb{C}} \Phi_\alpha(|z - w|)\, d\mu(w) \leq 2\pi \int_0^1 \Phi_\alpha(r)\, dr < \infty.$$

Theorem 2.3 now implies that $C_\alpha(K) > 0$. Thus, by definition, we also have $C_\alpha(E) > 0$.

If E is a Borel set of diameter at most one, we have

$$C_\beta(E) \leq k\, C_\alpha(E), \qquad (\beta \geq \alpha \geq 0),$$

where the constant $k > 0$ depends on the parameters α and β. Hence, for any Borel set E, the condition $C_\alpha(E) = 0$ implies $C_\beta(E) = 0$ for all $\beta \geq \alpha$. Based on this observation, the *Capacity dimension* of E is defined by

$$\dim_C(E) = \inf\{\, \alpha : C_\alpha(E) = 0 \,\}.$$

For more detailed treatment of this subject, see [44, Chap. 8].

2.3 Hausdorff Dimension

In the definition of a Hausdorff measure, we need a positive increasing and continuous function h which is defined on the interval $[0, \infty)$ and $h(0) = 0$. In particular, we will consider the family of functions

$$h_\alpha(t) = \begin{cases} t^\alpha & \text{if } \alpha > 0, \\ \frac{1}{\log 1/t} & \text{if } \alpha = 0. \end{cases} \qquad (2.8)$$

However, since, in our discussions, the behavior $h(t)$ for small values of t is important, sometimes we just give the definition of h on an interval $[0, \delta]$, where $\delta > 0$. The extension to (δ, ∞) is rather arbitrary.

Let E be any subset of $\mathbb{C}$. If there are open discs $B(z_n, r_n)$, $n \geq 1$, such that $E \subset \cup_n B(z_n, r_n)$, we say that $\cup_n B(z_n, r_n)$ is a covering of E. Fixing a function h as described above, define

$$\mu_{h,\delta}(E) = \inf \left\{ \sum_n h(r_n) : E \subset \bigcup_n B(z_n, r_n), \text{ and } r_n \leq \delta \right\},$$

where the infimum is taken over all possible coverings of E. The quantity $\mu_{h,\delta}(E)$ is a decreasing function of δ. Hence, we can also define

$$\mu_h(E) = \mu_{h,0}(E) = \lim_{\delta \to 0} \mu_{h,\delta}(E)$$

and

$$\mu_{h,\infty}(E) = \lim_{\delta \to \infty} \mu_{h,\delta}(E) = \inf \left\{ \sum_n h(r_n) : E \subset \bigcup_n B(z_n, r_n) \right\}.$$

Therefore, in evaluating $\mu_h(E)$ just the fine coverings of E are important, while in $\mu_{h,\infty}(E)$ all covering are considered. The functions $\mu_{h,\delta}$, where $0 \leq \delta \leq \infty$, are outer regular measures such that all Borel subsets of the complex plane are measurable with respect to them. In particular, these measures become more interesting when we exploit the family h_α. In this situation, we will write $\mu_{\alpha,\delta}$ and μ_α respectively for $\mu_{h_\alpha,\delta}$ and μ_{h_α}.

According to the definition above, we immediately see that

$$\mu_{h,\infty}(E) \leq \mu_{h,\delta}(E) \leq \mu_h(E).$$

Moreover, if $\mu_{h,\infty}(E) = 0$, then, given any $\varepsilon > 0$, there is a covering such that $E \subset \bigcup_n B(z_n, r_n)$ and $\sum_n h(r_n) < \varepsilon$. Therefore, at least $h(r_n) < \varepsilon$ which imposes the restriction

$$r_n < h^{-1}(\varepsilon), \qquad (n \geq 1). \qquad (2.9)$$

Back to the relation $E \subset \bigcup_n B(z_n, r_n)$, in the light of (2.9), we can now say $\mu_{h,h^{-1}(\varepsilon)}(E) < \varepsilon$. Let $\varepsilon \longrightarrow 0$ to get $\mu_h(E) = 0$. Therefore, we have

$$\mu_{h,\infty}(E) = 0 \iff \mu_h(E) = 0. \qquad (2.10)$$

The above definition also implies that $\mu_{h_1}(E) \leq C\mu_{h_2}(E)$ provided that the inequality $h_1(t) \leq C\,h_2(t)$ holds *at least* in a right neighborhood of the origin. This property immediately gives

$$0 \leq \mu_\alpha(E) < \infty \Longrightarrow \mu_\beta(E) = 0, \qquad (\beta > \alpha),$$

and

$$0 < \mu_\alpha(E) \leq \infty \Longrightarrow \mu_\beta(E) = \infty, \qquad (\beta < \alpha).$$

Hence, for a fixed Borel set E, the graph of $\alpha \longmapsto \mu_\alpha(E)$ is a step function with two possible values ∞ and 0 which breaks at some points in the interval $[0, \infty)$. At the breaking point, it can take a finite positive value. Based on this observation, the *Hausdorff dimension* of E is defined by

$$\dim_H(E) = \inf\{\,\alpha : \mu_\alpha(E) = 0\,\}.$$

If $\{\,\alpha : \mu_\alpha(E) = \infty\,\} \neq \emptyset$, we can also say

$$\dim_H(E) = \sup\{\,\alpha : \mu_\alpha(E) = \infty\,\}.$$

However, it is striking to know that

$$\dim_H(E) = \dim_C(E). \tag{2.11}$$

That is why we will simply write $\dim(E)$ to refer to the above common quantity.

For integers j, k and n, the set

$$Q = \left[\frac{j}{2^n}, \frac{j+1}{2^n}\right) \times \left[\frac{k}{2^n}, \frac{k+1}{2^n}\right)$$

is called a *dyadic square* in the complex plane. It has the side length $\ell(Q) = 2^{-n}$. If Q_1 and Q_2 are any two dyadic squares, it is fundamental to observe that if $Q_1 \cap Q_2 \neq \emptyset$, then either $Q_1 \subset Q_2$ or $Q_2 \subset Q_1$. We now define

$$m_h(E) = \inf\left\{\sum_n h(\ell(Q_n)) : E \subset \bigcup_n Q_n\right\},$$

where the infimum is taken over all dyadic covers of E. Since every dyadic square of side length r is contained in a disc of radius r and each disc of radius is contained in at most 25 dyadic squares Q with $r/2 < \ell(Q) \leq r$, we have

$$\mu_{h,\infty}(E) \leq m_h(E) \leq 25\,\mu_{h,\infty}(E). \tag{2.12}$$

Hence, according to (2.10) and (2.12), we deduce

$$\mu_{h,\infty}(E) = 0 \Longleftrightarrow \mu_h(E) = 0 \Longleftrightarrow m_h(E) = 0, \tag{2.13}$$

and this fact implies

$$
\begin{aligned}
\dim(E) &= \inf\{\,\alpha : \mu_\alpha(E) = 0\,\} \\
&= \inf\{\,\alpha : \mu_{\alpha,\infty}(E) = 0\,\} \\
&= \inf\{\,\alpha : m_\alpha(E) = 0\,\}.
\end{aligned}
$$

We now discuss the construction of a generalized Cantor set on the unit interval $[0,1]$. However, with small modifications, this method can also be applied for the unit circle $\mathbb{T}$. Let $E_0 = [0,1]$. Remove an open interval from the middle of E_0 to obtain E_1 which is the union of two closed intervals of length ℓ_1. In the next step, we remove an open interval from the middle of each interval in E_1 to obtain E_2 which is the union of four closed intervals of length ℓ_2. Hence, continuing this procedure, E_n would be the union of 2^n closed intervals of length ℓ_n. Put

$$
E = \bigcap_{n=0}^{\infty} E_n.
$$

The generalized Cantor set E is a perfect set and

$$
\mu_\alpha(E) = C \lim_{n \to \infty} 2^n \ell_n^\alpha, \tag{2.14}
$$

where the constant C just depends on α. Whenever needed, we can enumerate all the intervals that appear in this procedure, say I_n, $n \geq 1$, such that $I_{2^n}, I_{2^n+1}, \cdots, I_{2^{n+1}-1}$ represent the 2^n intervals that make E_n.

2.4 $\mathcal{E}_\varphi$ Has Logarithmic Capacity Zero

In Sect. 2.1, the exceptional set $\mathcal{E}_\varphi$ for an inner function φ was defined and we saw that this set has Lebesgue measure zero. Now, we have enough tools to refine this result by showing that its logarithmic capacity is in fact zero. Hence, in the light of Corollary 2.4, this result is a generalization of Theorem 2.1.

Theorem 2.5 (Frostman [23]) *Let φ be a nonconstant inner function for the open unit disc $\mathbb{D}$. Then the logarithmic capacity of $\mathcal{E}_\varphi$ is zero.*

Proof. Suppose, on the contrary, that the logarithmic capacity of $\mathcal{E}_\varphi$ is strictly positive, i.e. $C_0(\mathcal{E}_\varphi) > 0$. Hence, by Theorem 2.3, there is a positive Borel measure μ, with $\mu \neq 0$, whose support K lies in $\mathcal{E}_\varphi$ and the Green potential

$$
G_{0,\mu}(z) = \int_K \log\left|\frac{1 - \bar{w}z}{w - z}\right| d\mu(w), \qquad (z \in \mathbb{C}),
$$

is bounded above on $\mathbb{C}$. Clearly, we also have $G_{0,\mu} \geq 0$ on $\mathbb{D}$. The trick is to consider the function

$$\Phi(z) = (G_{0,\mu} \circ \varphi)(z) = \int_K \log \left| \frac{1 - \bar{w}\,\varphi(z)}{w - \varphi(z)} \right| d\mu(w), \qquad (z \in \mathbb{D}).$$

Since $K \subset \mathbb{D}$, the potential $G_{0,\mu}$ is continuous at all points of $\mathbb{T}$ (indeed, locally harmonic at such points) and, moreover, $G_{0,\mu} \equiv 0$ on $\mathbb{T}$. Hence, remembering that φ is inner, we deduce

$$\lim_{r \to 1} \Phi(re^{i\theta}) = \lim_{r \to 1} G_{0,\mu}(\varphi(re^{i\theta})) = G_{0,\mu}(\lim_{r \to 1} \varphi(re^{i\theta})) = 0$$

for almost all $e^{i\theta} \in \mathbb{T}$. Thus, by the dominated convergence theorem,

$$\lim_{r \to 1} \int_0^{2\pi} \Phi(re^{i\theta})\,d\theta = 0. \tag{2.15}$$

Note that we used the assumption that Φ is bounded on $\mathbb{D}$. But, by Fubini's theorem,

$$\int_0^{2\pi} \Phi(re^{i\theta})\,d\theta = \int_K \left(\int_0^{2\pi} \log \left| \frac{1 - \bar{w}\,\varphi(re^{i\theta})}{w - \varphi(re^{i\theta})} \right| d\theta \right) d\mu(w),$$

and thus, using the notation $M(r, w)$ exploited in the proof of Theorem 2.1, the relation (2.15) is rewritten as

$$\lim_{r \to 1} \int_K M(r, w)\,d\mu(w) = 0. \tag{2.16}$$

According to Theorem 1.14, we know that, for each fixed $w \in \mathbb{D}$,

$$\lim_{r \to 1} M(r, w) \tag{2.17}$$

exists and is positive. Moreover, by Fatou's lemma,

$$\int_0^{2\pi} \left(\liminf_{r \to 1} M(r, w) \right) d\mu(w) \leq \liminf_{r \to 1} \int_0^{2\pi} M(r, w)\,d\mu(w).$$

Therefore, by (2.16) and (2.17), we must have

$$\int_0^{2\pi} \left(\lim_{r \to 1} M(r, w) \right) d\mu(w) = 0.$$

This requires that

$$\lim_{r \to 1} M(r, w) = 0, \qquad (\mu\text{-a.e. } w \in K).$$

Hence, again by Theorem 1.14, φ_w is a Blaschke product for all $w \in K$, except possibly on a set of μ-measure zero. But the inclusion $K \subset \mathcal{E}_\varphi$ means that for no point of K, this assertion holds. This is a contradiction.

This version of Frostman's theorem shows that $\dim(\mathcal{E}_\varphi) = 0$.

2.5 The Cluster Set at a Boundary Point

Let $f \in H^\infty(\mathbb{D})$ and let $\zeta \in \mathbb{T}$. The Euclidean disc of radius r and center ζ is denoted by $D(\zeta, r)$. Then the *cluster set* of f at ζ is

$$\mathcal{C}(f, \zeta) = \bigcap_{r>0} \mathrm{Clos}_{\mathbb{C}}\, f\big(\mathbb{D} \cap D(\zeta, r)\big),$$

and its *range set* is defined by

$$\mathcal{R}(f, \zeta) = \bigcap_{r>0} f\big(\mathbb{D} \cap D(\zeta, r)\big).$$

Directly from these definitions, it is clear that $\mathcal{C}(f, \zeta)$ is a compact connected subset of $\mathbb{C}$ and

$$\mathcal{R}(f, \zeta) \subset \mathcal{C}(f, \zeta).$$

The finite intersection property of compact sets also reveals that $\mathcal{C}(f, \zeta)$ is nonempty. Moreover, f is continuous at ζ if and only if $\mathcal{C}(f, \zeta)$ is a singleton. But, $\mathcal{R}(f, \zeta)$ is a connected G_δ subset of $\mathbb{C}$, which also might be empty. For example, if f is not constant and has an analytic continuation across ζ, then, in the light of uniqueness theorem for analytic functions, we have

$$\mathcal{C}(f, \zeta) = \{f(\zeta)\} \qquad \text{and} \qquad \mathcal{R}(f, \zeta) = \emptyset.$$

As a matter of fact, the following topological interpretation helps us to verify the preceding claim and it is also useful in other applications. A point w belongs to $\mathcal{C}(f, \zeta)$ if and only if there is a sequence $(z_n)_{n \geq 1} \subset \mathbb{D}$ such that

$$\lim_{n \to \infty} z_n = \zeta \qquad \text{and} \qquad \lim_{n \to \infty} f(z_n) = w.$$

But, w belongs to $\mathcal{R}(f, \zeta)$ if and only if there is a sequence $(z_n)_{n \geq 1} \subset \mathbb{D}$ such that

$$\lim_{n \to \infty} z_n = \zeta \qquad \text{and} \qquad f(z_n) = w.$$

In other words, $\mathcal{R}(f, \zeta)$ is the set of values assumed by f infinitely many times in every neighborhood of ζ in $\mathbb{D}$, while, roughly speaking, $\mathcal{C}(f, \zeta)$ is the set of values that can be approximated by f in every neighborhood of ζ in $\mathbb{D}$.

Theorem 2.6 *Let φ be a nonconstant inner function, and let $\zeta \in \mathbb{T}$ be a singular point of φ. Then*

$$\mathcal{C}(\varphi, \zeta) = \overline{\mathbb{D}} \qquad and \qquad \mathbb{D} \setminus \mathcal{E}_\varphi \subset \mathcal{R}(\varphi, \zeta) \subset \mathbb{D}.$$

Proof. Since, by Theorem 2.5, $\mathcal{E}_\varphi$ has no interior and that, for an inner function, $\mathcal{C}(\varphi, \zeta)$ is necessarily a compact subset of $\overline{\mathbb{D}}$, the second assertion implies the first one.

We remind that the singular points of a Blaschke product are precisely the accumulation points of its zero set on $\mathbb{T}$. Now, if $w \in \mathbb{D} \setminus \mathcal{E}_\varphi$, then, by the definition of $\mathcal{E}_\varphi$,

$$\varphi_w(z) = (\tau_w \circ \varphi)(z) = \frac{w - \varphi(z)}{1 - \bar{w}\, \varphi(z)}, \qquad (z \in \mathbb{D}),$$

is a Blaschke product. Moreover, since φ cannot be analytically extended across ζ, then neither does φ_w. Otherwise, $\varphi = \tau_w \circ \varphi_w$ would have an analytic extension across ζ, which is a contradiction. In other words, ζ is also a singular point of φ_w. Thus, it is an accumulation point of the zeros of φ_w, which means that there is a sequence $(z_n)_{n \geq 1} \subset \mathbb{D}$ such that

$$\varphi_w(z_n) = \frac{w - \varphi(z_n)}{1 - \bar{w}\, \varphi(z_n)} = 0 \qquad and \qquad \lim_{n \to \infty} z_n = \zeta.$$

Therefore, $w \in \mathcal{R}(\varphi, \zeta)$.

The preceding result shows that a nonconstant inner function assumes all points of $\mathbb{D}$, except possibly a subset of logarithmic capacity zero, infinitely many times in each neighborhood of any of its singular points. Consider a Blaschke product whose zeros accumulates at all points of $\mathbb{T}$, or a singular inner functions constructed by a measure whose support is $\mathbb{T}$. Then such a function exhibits this rough behavior at all points of $\mathbb{T}$ and at the same time, according to Fatou's theorem, it has nontangential limits almost everywhere. This is a function with a very wild boundary behavior. Despite this wild behavior, still we can choose the zeros such that the nontangential limits exist at *all* points of $\mathbb{T}$. A comprehensive treatment of this subject is available at [15].

Chapter 3
The Derivative of Finite Blaschke Products

3.1 Elementary Formulas for B'

Let $w \in \mathbb{D}$. Consider the Blaschke factor

$$b(z) = \frac{w - z}{1 - \bar{w}\, z}.$$

Then direct computation shows that

$$b'(z) = -\frac{1 - |w|^2}{(1 - \bar{w}\, z)^2}. \tag{3.1}$$

We can easily obtain similar results for finite Blaschke products. In fact, put

$$B(z) = \prod_{k=1}^{n} \frac{z_k - z}{1 - \bar{z}_k\, z},$$

and

$$B_j(z) = \prod_{\substack{k=1 \\ k \neq j}}^{n} \frac{z_k - z}{1 - \bar{z}_k\, z}, \qquad (1 \leq j \leq n).$$

Then direct computation shows that

$$B'(z) = -\sum_{k=1}^{n} \frac{1 - |z_k|^2}{(1 - \bar{z}_k\, z)^2}\, B_k(z). \tag{3.2}$$

In particular, we have

$$B'(z_j) = -\frac{1}{1 - |z_j|^2} \prod_{\substack{k=1 \\ k \neq j}}^{n} \frac{z_k - z_j}{1 - \bar{z}_k\, z_j}, \qquad (1 \leq j \leq n). \tag{3.3}$$

J. Mashreghi, *Derivatives of Inner Functions*, Fields Institute Monographs 31, 39
DOI 10.1007/978-1-4614-5611-7_3, © Springer Science+Business Media New York 2013

Dividing both sides of (3.2) by B gives us the logarithmic derivative of B, i.e.

$$\frac{B'(z)}{B(z)} = \sum_{k=1}^{n} \frac{1 - |z_k|^2}{(1 - \bar{z}_k z)(z - z_k)}. \tag{3.4}$$

This relation is valid at all points of $\mathbb{C} \setminus \{z_k, 1/\bar{z}_k : 1 \leq k \leq n\}$. As a special case, at each $e^{i\theta} \in \mathbb{T}$, (3.4) is rewritten as

$$\frac{B'(e^{i\theta})}{e^{-i\theta} B(e^{i\theta})} = \sum_{k=1}^{n} \frac{1 - |z_k|^2}{|e^{i\theta} - z_k|^2}. \tag{3.5}$$

This identity immediately implies

$$|B'(e^{i\theta})| = \sum_{k=1}^{n} \frac{1 - |z_k|^2}{|e^{i\theta} - z_k|^2}. \tag{3.6}$$

The generalization of all above identities for infinite Blaschke products will be discussed in Sect. 4.1. The following result is a direct consequence of formula (3.6).

Lemma 3.1 *Let B be a finite Blaschke product for the open unit disc. Then*

$$B'(e^{i\theta}) \neq 0$$

for all $e^{i\theta} \in \mathbb{T}$.

3.2 The Cardinality of the Zeros of B'

If B is a finite Blaschke product of degree n, then, by (3.4), we see that $B' = P/Q$, where P and Q are polynomials and $\deg(P) \leq 2n - 2$. Lemma 3.1 ensures that no zeros of B' are of the unit circle $\mathbb{T}$. Hence, they are either in $\mathbb{D}$ or in $\mathbb{D}_e$. We explore further the locus of zeros of B' below and in Sect. 3.4.

Lemma 3.2 *Let B be finite Blaschke product for $\mathbb{D}$. Suppose that $w \in \mathbb{D} \setminus \{0\}$, and that w is not a zero of B. Then $B'(w) = 0$ if and only if $B'(1/\bar{w}) = 0$.*

Proof. For each $z \in \mathbb{D}$, $z \neq 0$, we have

$$B(z) \, \overline{B(1/\bar{z})} = 1.$$

Taking the derivative of both sides with respect to z gives

$$B'(z) \, \overline{B(1/\bar{z})} - \frac{1}{z^2} \, B(z) \, \overline{B'(1/\bar{z})} = 0.$$

This identity shows that

$$B'(w) = 0 \iff B'(1/\bar{w}) = 0.$$

One fact is implicitly stated in Lemma 3.2. If $z \in \mathbb{D}_e$ is such that $B'(z) = 0$, then certainly z is not a pole of B, which is equivalent to say that $w = 1/\bar{z}$ is not a zero of B. Hence, we must have $B'(1/\bar{z}) = 0$.

Theorem 3.3 *Let B be a finite Blaschke product of order n for $\mathbb{D}$. Write*

$$B(z) = e^{i\beta}\, z^{j_0} \prod_{k=1}^{m} \left(\frac{z_k - z}{1 - \bar{z}_k z} \right)^{j_k},$$

where $\beta \in \mathbb{R}$, j_k are some nonnegative integers with

$$j_0 + j_1 + \cdots + j_m = n,$$

and z_k's are distinct points in $\mathbb{D} \setminus \{0\}$. Then B' has precisely $n - 1$ solutions in $\mathbb{D}$ and the number of its solutions in $\mathbb{D}_e$ are

$$\begin{cases} = m & \text{if } j_0 \neq 0, \\ \leq m - 1 & \text{if } j_0 = 0. \end{cases}$$

Proof. To show that B' has always $n - 1$ zeros in $\mathbb{D}$, first suppose that the zeros are distinct and also neither B nor B' have any zeros at the origin. Hence, by (3.4), $B'(z) = 0$ if and only if

$$\sum_{k=1}^{n} \frac{1 - |z_k|^2}{(1 - \bar{z}_k z)(z - z_k)} = 0.$$

Multiplying both sides by

$$\prod_{k=1}^{n} (1 - \bar{z}_k z)(z - z_k),$$

we obtain a polynomial of degree $2n - 2$ with no from the set

$$\{0, z_1, \ldots, z_n, 1/\bar{z}_1, \ldots, 1/\bar{z}_n\}.$$

Therefore, by Lemma 3.2, there are exactly $n - 1$ zeros in $\mathbb{D}$ and $n - 1$ zeros in $\mathbb{D}_e$. In the general case, we can approximate B by a family B_ε of finite Blaschke products of order n with distinct zeros and such that neither B_ε nor B_ε' have any zeros at the origin. The convergence is uniform on compact subsets of $\mathbb{D}$. Hence, counting multiplicities, B' has exactly $n - 1$ zeros in $\mathbb{D}$. But, it might have fewer zeros in $\mathbb{D}_e$. The reason is that, in $\mathbb{D}_e$, the zeros of B_ε' may cluster at the poles of B or at ∞.

We now study the zeros of B' in $\mathbb{D}_e$. First assume that $j_0 \neq 0$. Then, by direct verification, we see that

$$B'(z) = z^{j_0 - 1} \times \frac{\prod_{k=1}^{m}(z - z_k)^{j_k - 1}}{\prod_{k=1}^{m}(z - 1/\bar{z}_k)^{j_k + 1}} \times P(z),$$

where P is a polynomial of degree $2m$ and it has no zeros among $\{0, z_1, \ldots, z_m\}$. Hence, B' has $n + m - 1$ zeros in $\mathbb{C}$. These are the zeros of B, repeated appropriately, and the zeros of P. For the latter, Lemma 3.2 applies and says its zeros are of the form

$$w_1, \ 1/\bar{w}_1, \ w_2, \ 1/\bar{w}_2, \ \ldots, \ w_m, \ 1/\bar{w}_m,$$

where $w_1, w_2, \ldots, w_m \in \mathbb{D} \setminus \{0, z_1, \ldots, z_m\}$.

Now, suppose that $j_0 = 0$. Write

$$B(z) = C \, \frac{\prod_{k=1}^{m}(z - z_k)^{j_k}}{\prod_{k=1}^{m}(z - 1/\bar{z}_k)^{j_k}} = C \left(1 + \frac{Q(z)}{\prod_{k=1}^{m}(z - 1/\bar{z}_k)^{j_k}} \right),$$

where C is a suitable constant and Q is a polynomial of degree $n - 1$. These representations show

$$B'(z) = \frac{\prod_{k=1}^{m}(z - z_k)^{j_k - 1}}{\prod_{k=1}^{m}(z - 1/\bar{z}_k)^{j_k + 1}} \times P(z),$$

where P is a polynomial of degree at most $2m - 2$ and it has no zeros from the set $\{z_1, \ldots, z_m\}$. Hence, B' has at most $n + m - 2$ zeros in $\mathbb{C}$. These are the zeros of B, repeated appropriately, and the zeros of P. In this case, P might have zeros at the origin. For the rest of its zeros, Lemma 3.2 applies. Therefore, P can have ℓ' zeros at the origin and the rest are of the form

$$w_1, \ 1/\bar{w}_1, \ w_2, \ 1/\bar{w}_2, \ \ldots, \ w_\ell, \ 1/\bar{w}_\ell,$$

where $w_1, w_2, \ldots, w_\ell \in \mathbb{D} \setminus \{0, z_1, \ldots, z_m\}$. Certainly $\ell' + 2\ell = \deg(P) \leq 2m - 2$. Hence, $\ell \leq m - 1$.

3.3 A Formula for $|B'|$

Let

$$B(z) = \frac{z_0 - z}{1 - \bar{z}_0 \, z}.$$

Then it is easy to verify that

$$|B(z)|^2 = 1 - \frac{(1 - |z|^2)(1 - |z_0|^2)}{|1 - \bar{z}_0 \, z|^2}. \tag{3.7}$$

This identity can be rewritten as

$$\frac{1 - |B(z)|^2}{1 - |z|^2} = \frac{1 - |z_0|^2}{|1 - \bar{z}_0\, z|^2},$$

and we can generalize it in the following manner.

Theorem 3.4 *Let*

$$B(z) = \prod_{j=1}^{n} \frac{z_j - z}{1 - \bar{z}_j\, z}.$$

Put $B_1 = 1$ and

$$B_k(z) = \prod_{j=1}^{k-1} \frac{z_j - z}{1 - \bar{z}_j\, z}, \qquad (2 \leq k \leq n).$$

Then, for each $z \in \mathbb{C} \setminus \mathbb{T}$,

$$\frac{1 - |B(z)|^2}{1 - |z|^2} = \sum_{k=1}^{n} |B_k(z)|^2 \frac{1 - |z_k|^2}{|1 - \bar{z}_k\, z|^2}.$$

Proof. The proof is by induction on the order of B. The case $n = 1$ is exactly the identity (3.7). Now, suppose that it holds for any Blaschke product of order $n - 1$. Hence, by induction hypothesis,

$$\frac{1 - |B_n(z)|^2}{1 - |z|^2} = \sum_{k=1}^{n-1} |B_k(z)|^2 \frac{1 - |z_k|^2}{|1 - \bar{z}_k\, z|^2}. \tag{3.8}$$

But,

$$B(z) = B_n(z) \frac{z_n - z}{1 - \bar{z}_n\, z},$$

and thus

$$1 - |B(z)|^2 = 1 - |B_n(z)|^2 \left| \frac{z_n - z}{1 - \bar{z}_n\, z} \right|^2$$

$$= 1 - |B_n(z)|^2 + |B_n(z)|^2 \left(1 - \left| \frac{z_n - z}{1 - \bar{z}_n\, z} \right|^2 \right)$$

$$= 1 - |B_n(z)|^2 + |B_n(z)|^2 \frac{(1 - |z|^2)(1 - |z_n|^2)}{|1 - \bar{z}_n\, z|^2}.$$

Hence, by (3.8),

$$\frac{1 - |B(z)|^2}{1 - |z|^2} = \frac{1 - |B_n(z)|^2}{1 - |z|^2} + |B_n(z)|^2 \frac{1 - |z_n|^2}{|1 - \bar{z}_n\, z|^2}$$

$$= \sum_{k=1}^{n} |B_k(z)|^2 \frac{1 - |z_k|^2}{|1 - \bar{z}_k\, z|^2}.$$

Appealing to the uniform convergence of partial Blaschke products on compact subsets of $\mathbb{D}$, with little effort we can generalize Theorem 3.4 for infinite Blaschke products. This formula will be needed in studying the derivatives of B.

Theorem 3.5 *Let*

$$B(z) = \prod_{j=1}^{\infty} \frac{\bar{z}_j}{z_j} \frac{z_j - z}{1 - \bar{z}_j z}.$$

Put $B_1 = 1$ and

$$B_k(z) = \prod_{j=1}^{k-1} \frac{z_j - z}{1 - \bar{z}_j z}, \qquad (k \geq 2).$$

Then, for each $z \in \mathbb{D}$,

$$\frac{1 - |B(z)|^2}{1 - |z|^2} = \sum_{k=1}^{\infty} |B_k(z)|^2 \frac{1 - |z_k|^2}{|1 - \bar{z}_k z|^2}.$$

Proof. Fix $z \in \mathbb{D}$. Then, on one hand, since $|B| \leq |B_n|$,

$$\frac{1 - |B(z)|^2}{1 - |z|^2} \geq \frac{1 - |B_n(z)|^2}{1 - |z|^2} = \sum_{k=1}^{n-1} |B_k(z)|^2 \frac{1 - |z_k|^2}{|1 - \bar{z}_k z|^2}.$$

Let $n \longrightarrow \infty$ to get

$$\frac{1 - |B(z)|^2}{1 - |z|^2} \geq \sum_{k=1}^{\infty} |B_k(z)|^2 \frac{1 - |z_k|^2}{|1 - \bar{z}_k z|^2}.$$

On the other hand, we have

$$\frac{1 - |B_n(z)|^2}{1 - |z|^2} \leq \sum_{k=1}^{\infty} |B_k(z)|^2 \frac{1 - |z_k|^2}{|1 - \bar{z}_k z|^2}.$$

This time, if we let $n \longrightarrow \infty$, Theorem 1.5 ensures that

$$\frac{1 - |B(z)|^2}{1 - |z|^2} \leq \sum_{k=1}^{\infty} |B_k(z)|^2 \frac{1 - |z_k|^2}{|1 - \bar{z}_k z|^2}.$$

The identity (3.6) gives the interesting formula

$$|B'(e^{i\theta})| = \sum_{k=1}^{n} \frac{1 - |z_k|^2}{|e^{i\theta} - z_k|^2}$$

for $|B'|$ on $\mathbb{T}$. Since B' is a continuous function on $\overline{\mathbb{D}}$, $|B'(z)|$ tends uniformly to $|B'(e^{i\theta})|$, as z approaches to the boundary $\mathbb{T}$. In other words,

$$\lim_{z \to e^{i\theta}} |B'(z)| = |B'(e^{i\theta})|$$

and the convergence is uniform with respect to θ. Now, we present a less trivial formula for $|B'(e^{i\theta})|$.

Theorem 3.6 *Let B be a finite Blaschke product. Then, for all $e^{i\theta} \in \mathbb{T}$,*

$$\lim_{z \to e^{i\theta}} \frac{1 - |B(z)|^2}{1 - |z|^2} = \lim_{z \to e^{i\theta}} \frac{1 - |B(z)|}{1 - |z|} = |B'(e^{i\theta})|,$$

and the convergence is uniform with respect to θ.

Proof. Let

$$B(z) = \prod_{j=1}^{n} \frac{z_j - z}{1 - \bar{z}_j z}.$$

Then, by Theorem 3.4, we have

$$\frac{1 - |B(z)|^2}{1 - |z|^2} = \sum_{k=1}^{n} |B_k(z)|^2 \frac{1 - |z_k|^2}{|1 - \bar{z}_k z|^2},$$

where $B_1 = 1$ and

$$B_k(z) = \prod_{j=1}^{k-1} \frac{z_j - z}{1 - \bar{z}_j z}, \qquad (2 \leq k \leq n).$$

Hence,

$$\lim_{z \to e^{i\theta}} \frac{1 - |B(z)|^2}{1 - |z|^2} = \sum_{k=1}^{n} \frac{1 - |z_k|^2}{|1 - \bar{z}_k e^{i\theta}|^2} = \sum_{k=1}^{n} \frac{1 - |z_k|^2}{|e^{i\theta} - z_k|^2},$$

and the convergence is uniform with respect to θ. Now, apply (3.6).

The identity

$$\lim_{z \to e^{i\theta}} \frac{1 - |B(z)|^2}{1 - |z|^2} = \lim_{z \to e^{i\theta}} \frac{1 - |B(z)|}{1 - |z|}$$

follows from the facts that $|B(z)| \longrightarrow 1$, as $|z| \longrightarrow 1$.

Corollary 3.7 *Let B be a finite Blaschke product. Then*

$$\lim_{|z| \to 1} \frac{|B'(z)| \, (1 - |z|^2)}{1 - |B(z)|^2} = 1,$$

and the convergence is uniform with respect to θ.

3.4 The Locus of the Zeros of B' in $\mathbb{D}$

There are various results about the relations between the zeros of a polynomial P and the zeros of its derivatives. The oldest goes back to Gauss and Lucas [31]. This result affirms that the zeros of P' are in the convex hull of the zeros of P. The success of this method is based on the following observation. If we decompose P as

$$P(z) = C\,(z - z_1)^{k_1} \cdots (z - z_n)^{k_n},$$

then

$$\frac{P'(z)}{P(z)} = \frac{k_1}{z - z_n} + \cdots + \frac{k_n}{z - z_n}.$$

The last representation enables us to locate the zeros of P'.

Adopting the classical definition from the Euclidean geometry, we say that a set $A \subset \mathbb{D}$ is hyperbolically convex if

$$z_1, z_2 \in A \implies \forall t \in [0,1],\ \frac{z_1 - \frac{z_1 - z_2}{1 - \bar{z}_1 z_2}\,t}{1 - \bar{z}_1 \frac{z_1 - z_2}{1 - \bar{z}_1 z_2}\,t} \in A.$$

Then the hyperbolic convex hull of a set $A \subset \mathbb{D}$ is the smallest hyperbolic convex set which contains A. This is clearly the intersection of all hyperbolic convex sets that contain A. If we need the closed hyperbolic convex hull, we must consider the intersection of all closed hyperbolic convex sets that contain A. We remind that $\mathbb{D}_- = \mathbb{D} \cap \{z : \Im z < 0\}$ and $\mathbb{D}_+ = \mathbb{D} \cap \{z : \Im z > 0\}$. The following result is due to J. Walsh [46].

Theorem 3.8 *Let B be a finite Blaschke product. Then the zeros of B' which are inside $\mathbb{D}$ are in the hyperbolic convex hull of the zeros of B.*

Proof. First suppose that all zeros of B are in $\mathbb{D}_+$. Then, by (3.4),

$$\Im\left(\frac{B'(z)}{B(z)}\right) = \sum_{k=1}^{n} \Im\left(\frac{1 - |z_k|^2}{(1 - \bar{z}_k\,z)(z - z_k)}\right).$$

Put

$$\varphi(z) = \frac{1 - |a|^2}{(1 - \bar{a}\,z)(z - a)},$$

where a, with $\Im a > 0$ and $|a| < 1$, is fixed. We need to obtain the image of $\mathbb{D}_-$ under φ. To do so, we study the image of the boundary of $\partial \mathbb{D}_-$ under φ. On the lower semicircle

$$\mathbb{T}_- = \{e^{i\theta} : -\pi \leq \theta \leq 0\}$$

we have

$$\varphi(e^{i\theta}) = \frac{1 - |a|^2}{(1 - \bar{a}\, e^{i\theta})(e^{i\theta} - a)} = \frac{1 - |a|^2}{|e^{i\theta} - a|^2}\, e^{-i\theta}.$$

Therefore, $\mathbb{T}_-$ is mapped to an arc in $\mathbb{C}_+ = \{z : \Im z > 0\}$. Moreover, on the interval $t \in (-1, 1)$, we have

$$\varphi(t) = \frac{1 - |a|^2}{(1 - \bar{a}\,t)(t - a)} = \frac{1 - |a|^2}{|t - a|^2} \times \frac{t - \bar{a}}{1 - \bar{a}\,t} = \frac{1 - |a|^2}{|t - a|^2}\, \tau_t(\bar{a}).$$

Thus $(-1, 1)$ is also mapped to an arc in $\mathbb{C}_+$. In short, $\partial \mathbb{D}_-$ is mapped to a closed arc in $\mathbb{C}_+$. Therefore, we deduce that φ maps $\mathbb{D}_-$ into $\mathbb{C}_+$. This fact implies that B' has no zeros in $\mathbb{D}_-$. By continuity, if all zeros of B are in $\overline{\mathbb{D}}_+$, then all the zeros of B' which are in the open unit disc are necessarily in $\overline{\mathbb{D}}_+$.

Let $f = B \circ \tau_a$. Then f is also a finite Blaschke product with zeros $\tau_a(z_1), \tau_a(z_2), \ldots, \tau_a(z_n)$. Moreover, if we denote the zeros of B' in $\mathbb{D}$ by $w_1, w_2, \ldots, w_{n-1}$, the zeros of f' in $\mathbb{D}$ would be $\tau_a(w_1), \tau_a(w_2), \ldots, \tau_a(w_{n-1})$.

If we choose a such that

$$\Im \tau_a(z_k) \geq 0, \qquad (1 \leq k \leq n),$$

then the preceding observation shows that

$$\Im \tau_a(w_k) \geq 0, \qquad (1 \leq k \leq n - 1).$$

Applying again the transformation τ_a, we see that if the zeros of B are on one side of the hyperbolic line

$$\frac{a - z}{1 - \bar{a}\,z} = t, \qquad t \in [-1, 1],$$

then the zeros of B' are also on the same side. Similar comments apply if we replace τ_a by a rotation ρ_γ. The intersection of all such lines gives the hyperbolic convex hull of the zeros of B.

Remark. The argument above also works for infinite Blaschke products.

Let $a, b \in \mathbb{D}$ and put

$$B(z) = \left(\frac{a - z}{1 - \bar{a}\,z}\right)^m \left(\frac{b - z}{1 - \bar{b}\,z}\right)^n.$$

Clearly B' has $m + n - 1$ zeros in $\mathbb{D}$ which are a, $m - 1$ times, and b, $n - 1$ times, and the last one c which is the solution of the equation

$$\frac{m(1 - |a|^2)}{(1 - \bar{a}\,z)^2} \left(\frac{a - z}{1 - \bar{a}\,z}\right)^{m-1} \left(\frac{b - z}{1 - \bar{b}\,z}\right)^n + \left(\frac{a - z}{1 - \bar{a}\,z}\right)^m \left(\frac{b - z}{1 - \bar{b}\,z}\right)^{n-1} \frac{n(1 - |b|^2)}{(1 - \bar{b}\,z)^2} = 0.$$

Rewrite this equation as

$$\left(\frac{z-a}{1-a\,\bar z}\right)\Big/\left(\frac{z-b}{1-b\,\bar z}\right) = -\left(\frac{m(1-|a|^2)}{|1-\bar a\,z|^2}\right)\Big/\left(\frac{n(1-|b|^2)}{|1-\bar b\,z|^2}\right).$$

This identity reveals that a, b, c are on the same hyperbolic line. Moreover, as m and n are any arbitrary positive integers, the point c traverses a dense subset of the hyperbolic line segment between a and b.

3.5 B Has a Nonzero Residue

The finite Blaschke product $B(z) = z^m$, $m \geq 0$, is an entire function. All other Blaschke products have *finite poles* and thus we can talk about their residues. That is why in the following theorem we assume that B has a finite pole.

Theorem 3.9 (Heins [27]) *Let B be a finite Blaschke product having at least one finite pole. Then B has a nonzero residue.*

Proof. Let

$$B(z) = e^{i\beta}\, z^m \prod_{n=1}^{N}\left(\frac{z-z_n}{1-\bar z_n\, z}\right)^{m_n}. \tag{3.9}$$

By assumption, at least one non-zero z_n exists. Since B is analytic on $\overline{\mathbb{D}}$ it has a primitive, say $\mathbb{B}$, in an open disc containing $\overline{\mathbb{D}}$. Furthermore, since $\mathbb{B}$ involves an arbitrary constant, we choose the constant such that $\mathbb{B}(0) = 0$. Thus, by the fundamental theorem of calculus, for each $z \in \overline{\mathbb{D}}$,

$$\mathbb{B}(z) = \int_0^z \mathbb{B}'(\zeta)\, d\zeta = \int_0^z B(\zeta)\, d\zeta. \tag{3.10}$$

Since $\mathbb{B}'$ is analytic on $\overline{\mathbb{D}}$, the integral in (3.10) is independent of the path of integration. This representation enables us to estimate $\mathbb{B}$ on the unit circle $\mathbb{T}$. By (3.9), for each $e^{i\theta} \in \mathbb{T}$,

$$\mathbb{B}(e^{i\theta}) = \int_0^{e^{i\theta}} B(z)\, dz$$

$$= \int_0^1 e^{i\beta}\, r^m e^{im\theta} \prod_{n=1}^{N}\left(\frac{re^{i\theta}-z_n}{1-\bar z_n\, re^{i\theta}}\right)^{m_n} e^{i\theta}\, dr. \tag{3.11}$$

The function

$$\prod_{n=1}^{N}\left(\frac{re^{i\theta}-z_n}{1-\bar z_n\, re^{i\theta}}\right)^{m_n}$$

is also a finite Blaschke product. Thus, inside the unit disc $\mathbb{D}$, it is bounded by one. Therefore, by (3.11),

$$
\begin{aligned}
|\mathbb{B}(e^{i\theta})| &= \left| \int_0^1 e^{i\beta}\, r^m\, e^{im\theta} \prod_{n=1}^N \left(\frac{re^{i\theta} - z_n}{1 - \bar{z}_n\, re^{i\theta}} \right)^{m_n} e^{i\theta}\, dr \right| \\
&\leqslant \int_0^1 r^m\, dr = \frac{1}{m+1}
\end{aligned}
\tag{3.12}
$$

for each $e^{i\theta} \in \mathbb{T}$.

Applying partial fraction expansion to (3.9), we have

$$
B(z) = \sum_{n=0}^m \alpha_n\, z^n + \sum_{n=1}^N \sum_{\ell=1}^{m_n} \frac{\beta_{n,\ell}}{(1 - \bar{z}_n z)^\ell}.
\tag{3.13}
$$

Suppose that all residues of B, i.e. all $\beta_{n,1}$ $(1 \leqslant n \leqslant N)$, are zero. Then

$$
\mathbb{B}(z) = \sum_{n=0}^m \frac{\alpha_n}{n+1}\, z^{n+1} + \alpha + \sum_{n=1}^N \sum_{\ell=2}^{m_n} \frac{\left(\frac{\beta_{n,\ell}}{\bar{z}_n\,(\ell-1)} \right)}{(1 - \bar{z}_n z)^{\ell-1}}
\tag{3.14}
$$

is a primitive of B on $\mathbb{C} \setminus \{1/\bar{z}_1, \cdots, 1/\bar{z}_N\}$. The constant α is arbitrary and, as we agreed, we choose it such that $\mathbb{B}(0) = 0$. Since B has a zero of order m at origin, $\mathbb{B}(0) = 0$, and $\mathbb{B}' = B$, then $\mathbb{B}$ has a zero of order $m+1$ at origin. Thus, by taking the common denominator in (3.14),

$$
\mathbb{B}(z) = \frac{z^{m+1}\, P(z)}{\prod_{n=1}^N (1 - \bar{z}_n z)^{m_n - 1}},
\tag{3.15}
$$

where P is a polynomial of degree at most $\sum_{n=1}^N (m_n - 1)$.

On the other hand, a direct consequence of (3.13) and (3.14) is that

$$
\lim_{z \to \infty} \frac{(m+1)\,\mathbb{B}(z)}{z\, B(z)} = 1.
\tag{3.16}
$$

To end the proof, consider the function

$$
f(z) = \frac{(m+1)\,\mathbb{B}(z)}{z\, B(z)}.
$$

According to (3.15) and (3.9),

$$
f(z) = \frac{(m+1)\, P(z) \prod_{n=1}^N (1 - \bar{z}_n z)}{\prod_{n=1}^N (z - z_n)^{m_n}}.
$$

Hence, f is analytic on $\mathbb{C} \setminus \mathbb{D}$. We assumed that B has finite poles. Thus f has at least *one zero* in $\mathbb{C} \setminus \overline{\mathbb{D}}$.

Now, we are able to find the contradiction. By (3.16),

$$\lim_{z \to \infty} f(z) = 1,$$

and, according to (3.12),

$$|f(e^{i\theta})| = \left| \frac{(m+1)\,\mathbb{B}(e^{i\theta})}{e^{i\theta}\,B(e^{i\theta})} \right| = \left| (m+1)\,\mathbb{B}(e^{i\theta}) \right| \leqslant 1$$

for each $e^{i\theta} \in \mathbb{T}$. But, the maximum principle ensures that if f is analytic on $\mathbb{C} \setminus \mathbb{D}$, and $|f(\zeta)| \leqslant 1$, for each $\zeta \in \mathbb{T}$, and $\lim_{z \to \infty} f(z) = 1$, then $f \equiv 1$. This contradicts that f has zeros in $\mathbb{C} \setminus \mathbb{D}$. Therefore, our hypothesis is wrong, i.e. there is an integer n, with $1 \leqslant n \leqslant N$, such that $\beta_{n,1} \neq 0$.

For further discussion on this result, see [32].

Chapter 4
Angular Derivative

4.1 Elementary Formulas for B' and S'

The canonical factorization theorem says that any inner function ϕ can be decomposed as $\phi = BS$, where

$$B(z) = \gamma \prod_{n=1}^{\infty} \frac{|z_n|}{z_n} \frac{z_n - z}{1 - \bar{z}_n z}$$

is the Blaschke product formed with the zeros of ϕ in $\mathbb{D}$, and

$$S(z) = \exp\left(-\int_{\mathbb{T}} \frac{e^{it} + z}{e^{it} - z} \, d\sigma(e^{it}) \right)$$

is the singular inner function formed with the finite positive singular Borel measure σ. In this section, we find some elementary formulas for the derivative of B and S.

Let

$$B_k(z) = \prod_{n=1, \, n \neq k}^{\infty} \frac{|z_n|}{z_n} \frac{z_n - z}{1 - \bar{z}_n z}, \qquad (k \geq 1).$$

Thus, $B_k(z_n) = 0$, if $n \neq k$, and

$$B_k(z_k) = \prod_{n=1, \, n \neq k}^{\infty} \frac{|z_n|}{z_n} \frac{z_n - z_k}{1 - \bar{z}_n z_k}, \qquad (k \geq 1). \tag{4.1}$$

According to Theorem 1.5, the Blaschke product B converges uniformly on compact subsets of

$$\Omega = \mathbb{C} \setminus \{1/\bar{z}_n : n \geq 1\}^{cl}.$$

Hence, at each point of Ω, we can evaluate the derivative term by term to obtain

J. Mashreghi, *Derivatives of Inner Functions*, Fields Institute Monographs 31, 51
DOI 10.1007/978-1-4614-5611-7_4, © Springer Science+Business Media New York 2013

$$B'(z) = -\sum_{n=1}^{\infty} \frac{|z_n|}{z_n} \frac{1 - |z_n|^2}{(1 - \bar{z}_n z)^2} B_n(z). \tag{4.2}$$

In particular, by (4.1), we have

$$B'(z_k) = -\frac{|z_k|}{z_k} \frac{1}{1 - |z_k|^2} \prod_{n=1, n \neq k}^{\infty} \frac{|z_n|}{z_n} \frac{z_n - z_k}{1 - \bar{z}_n z_k}, \qquad (k \geq 1). \tag{4.3}$$

In our applications, we need the special formula

$$|B'(z_k)| = \frac{1}{1 - |z_k|^2} \prod_{n=1, n \neq k}^{\infty} \left| \frac{z_n - z_k}{1 - \bar{z}_n z_k} \right|, \qquad (k \geq 1), \tag{4.4}$$

which implies the interesting estimation

$$\frac{\delta}{2(1 - |z_k|)} \leq |B'(z_k)| \leq \frac{1}{1 - |z_k|}, \qquad (k \geq 1), \tag{4.5}$$

where

$$\delta = \inf_{k \geq 1} \prod_{n=1, n \neq k}^{\infty} \left| \frac{z_n - z_k}{1 - \bar{z}_n z_k} \right|,$$

for an interpolating Blaschke sequence.

The sum in (4.2) is uniformly convergent on compact subsets of Ω. We also recall that $\mathbb{D} \subset \Omega$, and thus (4.2) is valid at all points of $\mathbb{D}$ and the series converges uniformly on compact subsets of $\mathbb{D}$. Dividing both sides of (4.2) by B gives the logarithmic derivative formula

$$\frac{B'(z)}{B(z)} = -\sum_{n=1}^{\infty} \frac{1 - |z_n|^2}{(1 - \bar{z}_n z)(z_n - z)}. \tag{4.6}$$

Note that this relation is valid at all points of $\mathbb{C} \setminus \{z_n, 1/\bar{z}_n : n \geq 1\}^{cl}$ and, moreover, the series is uniformly convergent on compact subsets of this open set.

If $e^{i\theta} \in \mathbb{T}$ is not an accumulation point of the sequence $(z_n)_{n \geq 1}$, then B is analytic at $e^{i\theta}$ and (4.6) is rewritten as

$$\frac{B'(e^{i\theta})}{e^{-i\theta} B(e^{i\theta})} = \sum_{n=1}^{\infty} \frac{1 - |z_n|^2}{|e^{i\theta} - z_n|^2}. \tag{4.7}$$

This identity immediately implies the interesting formula

$$|B'(e^{i\theta})| = \sum_{n=1}^{\infty} \frac{1 - |z_n|^2}{|e^{i\theta} - z_n|^2} \tag{4.8}$$

at all points of the unit circle where B is analytic. This is a generalization of (3.6).

With a little effort, similar formulas can be obtained for S'. First of all, we have

$$\frac{S'(z)}{S(z)} = -\int_{\mathbb{T}} \frac{2e^{it}}{(e^{it} - z)^2} \, d\sigma(e^{it}) \tag{4.9}$$

for all z in the complex plane except the support of σ. The rough estimation

$$\left| \frac{S'(z)}{S(z)} \right| \le \int_{\mathbb{T}} \frac{2d\sigma(e^{it})}{|e^{it} - z|^2}, \qquad (z \in \mathbb{D}), \tag{4.10}$$

will be useful in several applications. If $e^{i\theta} \in \mathbb{T} \setminus \text{supp}\, \sigma$, then the formula (4.9) reduces to

$$\frac{S'(e^{i\theta})}{e^{-i\theta} S(e^{i\theta})} = \int_{\mathbb{T}} \frac{d\sigma(e^{it})}{2\sin^2(\frac{t-\theta}{2})} = \int_{\mathbb{T}} \frac{2d\sigma(e^{it})}{|e^{i\theta} - e^{it}|^2}. \tag{4.11}$$

Hence, at such points, we have

$$|S'(e^{i\theta})| = \int_{\mathbb{T}} \frac{d\sigma(e^{it})}{2\sin^2(\frac{t-\theta}{2})} = \int_{\mathbb{T}} \frac{2d\sigma(e^{it})}{|e^{i\theta} - e^{it}|^2}. \tag{4.12}$$

Since $\phi = BS$, we have

$$\frac{\phi'}{\phi} = \frac{B'}{B} + \frac{S'}{S}$$

at all points of $\mathbb{C}$, except the zeros and poles of B, their accumulation points on $\mathbb{T}$, and the support of σ. Hence, in the light of (4.6) and (4.9), at such points we have

$$\frac{\phi'(z)}{\phi(z)} = -\sum_{n=1}^{\infty} \frac{1 - |z_n|^2}{(1 - \bar{z}_n z)(z_n - z)} - \int_{\mathbb{T}} \frac{2e^{it}}{(e^{it} - z)^2} \, d\sigma(e^{it}). \tag{4.13}$$

In particular, at points $e^{i\theta} \in \mathbb{T}$ which are away from the support of σ and the accumulation points of the zeros of B, by (4.7) and (4.11), we obtain

$$\frac{\phi'(e^{i\theta})}{e^{-i\theta} \phi(e^{i\theta})} = \sum_{n=1}^{\infty} \frac{1 - |z_n|^2}{|e^{i\theta} - z_n|^2} + \int_{\mathbb{T}} \frac{2d\sigma(e^{it})}{|e^{i\theta} - e^{it}|^2}. \tag{4.14}$$

Therefore, at such points where ϕ is analytic, we have

$$|\phi'(e^{i\theta})| = \sum_{n=1}^{\infty} \frac{1 - |z_n|^2}{|e^{i\theta} - z_n|^2} + \int_{\mathbb{T}} \frac{2d\sigma(e^{it})}{|e^{i\theta} - e^{it}|^2}. \tag{4.15}$$

With appropriate definitions and conventions, the above formula actually makes sense at all points of $\mathbb{T}$.

In establishing formulas (4.9), (4.11) and (4.12), we did not use the fact that σ is singular with respect to the Lebesgue measure. Hence, these formulas hold for a wider class of functions. To clarify this claim, note that an arbitrary element $f \in H^1(\mathbb{D})$ has the representation

$$f(z) = \prod_{n=1}^{\infty} \frac{|z_n|}{z_n} \frac{z_n - z}{1 - \bar{z}_n z} \times \exp\left(-\int_{\mathbb{T}} \frac{e^{it} + z}{e^{it} - z} \, d\mu(e^{it})\right)$$

where the Blaschke product is formed with the zeros of f and the measure μ has the Lebesgue decomposition

$$d\mu(e^{it}) = -\log|f(e^{it})| \frac{dt}{2\pi} + d\sigma(e^{it}),$$

Hence, the formulas (4.13) and (4.14) also work for any $f \in H^1$, and we obtain

$$\frac{f'(z)}{f(z)} = -\sum_{n=1}^{\infty} \frac{1 - |z_n|^2}{(1 - \bar{z}_n z)(z_n - z)} - \int_{\mathbb{T}} \frac{2e^{it}}{(e^{it} - z)^2} \, d\mu(e^{it}),$$

and

$$\frac{f'(e^{i\theta})}{e^{-i\theta} f(e^{i\theta})} = \sum_{n=1}^{\infty} \frac{1 - |z_n|^2}{|e^{i\theta} - z_n|^2} + \int_{\mathbb{T}} \frac{2d\mu(e^{it})}{|e^{i\theta} - e^{it}|^2}.$$

Using (4.2) and (4.9), we also get the rough estimations

$$|B'(z)| \leq \sum_{n=1}^{\infty} \frac{1 - |z_n|^2}{|1 - \bar{z}_n z|^2}, \qquad (z \in \mathbb{D}), \tag{4.16}$$

and

$$|S'(z)| \leq \int_{\mathbb{T}} \frac{d\sigma(e^{it})}{|e^{it} - z|^2}, \qquad (z \in \mathbb{D}). \tag{4.17}$$

From another point of view, by the Schwarz–Pick theorem, we also have

$$|\phi'(z)| \leq \frac{1 - |\phi(z)|^2}{1 - |z|^2}, \qquad (z \in \mathbb{D}), \tag{4.18}$$

for any inner function ϕ. These two relations will be needed later on.

4.2 Some Estimations for H^p-Means

For each $re^{i\theta} \in \mathbb{D}$, $|\theta| \leq \pi$, we have

$$|1 - re^{i\theta}|^2 = 1 + r^2 - 2r\cos\theta$$
$$= (1 - r)^2 + 4r\sin^2(\theta/2).$$

Hence, on the one hand,

$$|1 - re^{i\theta}|^2 \le (1 - r)^2 + |\theta|^2 \tag{4.19}$$

and, on the other hand, at least for $r \ge 1/4$,

$$|1 - re^{i\theta}|^2 \ge (1 - r)^2 + |\theta/\pi|^2.$$

Therefore, for all $re^{i\theta} \in \mathbb{D}$,

$$|1 - re^{i\theta}|^2 \ge \frac{(1-r)^2 + |\theta|^2}{22}. \tag{4.20}$$

Surely, 22 is not the best constant. But, this is not important for our applications. In short, instead of (4.19) and (4.20), we can write

$$|1 - re^{i\theta}|^p \asymp (1 - r)^p + |\theta|^p, \tag{4.21}$$

where the constants involved just depend on the parameter $p \in (0, \infty)$. the notion $f \asymp g$ means that there are two positive constants c and C such that

$$c\,|g(x)| \le |f(x)| \le C\,|g(x)|$$

for all x in the domain of definition of f and g.

Lemma 4.1 *Let $w \in \mathbb{D}$, and let $0 < p < \infty$. Then, as $|w| \longrightarrow 1$,*

$$\int_0^{2\pi} \frac{d\theta}{|1 - \bar{w}\,e^{i\theta}|^p} \asymp \begin{cases} 1 & \text{if } 0 < p < 1, \\ -\log(1 - |w|) & \text{if } p = 1, \\ \dfrac{1}{(1 - |w|)^{p-1}} & \text{if } p > 1. \end{cases}$$

The constants involved just depend on the parameter p.

Proof. A change of variable shows

$$\int_0^{2\pi} \frac{d\theta}{|1 - \bar{w}\,e^{i\theta}|^p} = 2\int_0^{\pi} \frac{d\theta}{|1 - |w|\,e^{i\theta}|^p}.$$

Hence, by (4.21),

$$\int_0^{2\pi} \frac{d\theta}{|1 - \bar{w}\,e^{i\theta}|^p} \asymp \int_0^{\pi} \frac{d\theta}{(1 - |w|)^p + \theta^p}.$$

If $0 < p < 1$, the inequalities

$$\frac{1}{1 + \pi^p} \leq \frac{1}{(1 - |w|)^p + \theta^p} \leq \frac{1}{\theta^p}$$

show that

$$\int_0^{2\pi} \frac{d\theta}{|1 - \bar{w}\, e^{i\theta}|^p} \asymp 1.$$

If $p = 1$, then

$$\int_0^{2\pi} \frac{d\theta}{|1 - \bar{w}\, e^{i\theta}|^p} \asymp \int_0^{\pi} \frac{d\theta}{(1 - |w|) + \theta}$$
$$= \log\big((1 - |w|) + \pi\big) - \log(1 - |w|)$$
$$\asymp \log \frac{1}{(1 - |w|)}.$$

Finally, if $p > 1$, then

$$\int_0^{2\pi} \frac{d\theta}{|1 - \bar{w}\, e^{i\theta}|^p} \asymp \int_0^{\pi} \frac{d\theta}{(1 - |w|)^p + \theta^p}$$
$$= \int_0^{1-|w|} \frac{d\theta}{(1 - |w|)^p + \theta^p} + \int_{1-|w|}^{\pi} \frac{d\theta}{(1 - |w|)^p + \theta^p}$$
$$\asymp \int_0^{1-|w|} \frac{d\theta}{(1 - |w|)^p} + \int_{1-|w|}^{\pi} \frac{d\theta}{\theta^p}$$
$$\asymp \frac{1}{(1 - |w|)^{p-1}}.$$

The estimations obtained above allow us to introduce a family of analytic functions with close ties to Hardy spaces.

Corollary 4.2 *Let $0 < \alpha < \infty$, and define*

$$f_\alpha(z) = \frac{1}{(1 - z)^\alpha}, \qquad (z \in \mathbb{D}).$$

Then

$$f_\alpha \in \bigcap_{0 < p < \frac{1}{\alpha}} H^p(\mathbb{D}),$$

but

$$f_\alpha \notin H^{\frac{1}{\alpha}}(\mathbb{D}).$$

Let $w \in \mathbb{D}$, and write

$$b_w(z) = \frac{|w|}{w} \frac{w - z}{1 - \bar{w}\, z}, \qquad (z \in \mathbb{D}).$$

Hence,

$$\int_0^{2\pi} |b_w'(re^{i\theta})|^p \, d\theta = \int_0^{2\pi} \frac{(1-|w|^2)^p}{|1-\bar{w}\,re^{i\theta}|^{2p}} \, d\theta$$

and, by Lemma 4.1,

$$\int_0^{2\pi} |b_w'(re^{i\theta})|^p \, d\theta \asymp \begin{cases} (1-|w|)^p & \text{if } 0 < p < \frac{1}{2}, \\[2mm] (1-|w|)^{\frac{1}{2}} \log \dfrac{1}{(1-|rw|)} & \text{if } \quad p = \frac{1}{2}, \\[2mm] \dfrac{(1-|w|)^p}{(1-|wr|)^{2p-1}} & \text{if } \quad p > \frac{1}{2}. \end{cases} \qquad (4.22)$$

In each case, the constants involved just depend on p. More specifically, they neither depend on r nor on w. Therefore, letting $r \longrightarrow 1$, we obtain

$$\|b_w'\|_p \asymp \begin{cases} (1-|w|) & \text{if } 0 < p < \frac{1}{2}, \\[2mm] (1-|w|) \log^2 \dfrac{1}{(1-|w|)} & \text{if } \quad p = \frac{1}{2}, \\[2mm] (1-|w|)^{\frac{1-p}{p}} & \text{if } \quad p > \frac{1}{2}. \end{cases} \qquad (4.23)$$

4.3 Some Estimations for A^p-Means

Parallel to the results obtained in Sect. 4.2, we continue and obtain some similar estimations for the A^p-means.

Lemma 4.3 *Let $w \in \mathbb{D}$, let $q > -1$, and let $0 < p < \infty$. Then, as $|w| \longrightarrow 1$,*

$$\int_0^{2\pi} \int_0^1 \frac{(1-r)^q}{|1-\bar{w}\,re^{i\theta}|^p} \, r\,dr\,d\theta \asymp \begin{cases} 1 & \text{if } 0 < p < q+2, \\[2mm] -\log(1-|w|) & \text{if } \quad p = q+2, \\[2mm] \dfrac{1}{(1-|w|)^{p-q-2}} & \text{if } \quad p > q+2. \end{cases}$$

Proof. By the generalized binomial expansion formula, we have

$$\frac{1}{(1-z)^{\frac{p}{2}}} = \sum_{n=0}^{\infty} \frac{\Gamma(n+\frac{p}{2})}{n!\,\Gamma(\frac{p}{2})} \, z^n, \qquad (z \in \mathbb{D}).$$

Write

$$I_{p,q}(w) = \int_0^{2\pi} \int_0^1 \frac{(1-r)^q}{|1-\bar{w}\,re^{i\theta}|^p} \, r\,dr\,d\theta.$$

Hence,

$$I_{p,q}(w) = \sum_{n=0}^{\infty} \frac{\Gamma^2(n + \frac{p}{2})}{(n!)^2 \, \Gamma^2(\frac{p}{2})} \, |w|^{2n} \int_0^{2\pi} \int_0^1 r^{2n}(1-r)^q \, rdrd\theta$$

$$= 2\pi \frac{\Gamma(q+1)}{\Gamma^2(\frac{p}{2})} \sum_{n=0}^{\infty} \frac{\Gamma^2(n + \frac{p}{2}) \, (2n+1)!}{(n!)^2 \, \Gamma(2n+q+3)} \, |w|^{2n}.$$

According to Stirling's formula,

$$\frac{\Gamma^2(n + \frac{p}{2}) \, (2n+1)!}{(n!)^2 \, \Gamma(2n+q+3)} \asymp n^{p-q-3}$$

as $n \longrightarrow \infty$.

Case I: If $0 < p < q+2$, the function

$$\sum_{n=1}^{\infty} \frac{|w|^{2n}}{n^{q-p+3}}$$

defines a bounded function on $\mathbb{D}$, and thus $I_{p,q}(w) \asymp 1$.

Case II: If $p = q+2$, then

$$\sum_{n=1}^{\infty} \frac{|w|^{2n}}{n} \asymp \log \frac{1}{(1-|w|)}$$

and thus $I_{p,q}(w) \asymp \log \frac{1}{(1-|w|)}$.

Case III: Finally, if $p > q+2$, then

$$\sum_{n=1}^{\infty} \frac{|w|^{2n}}{n^{q+3-p}} \asymp \frac{1}{(1-|w|)^{p-q-2}}$$

and thus $I_{p,q}(w) \asymp \frac{1}{(1-|w|)^{p-q-2}}$. This last fact comes from the expansion

$$\frac{1}{(1-z)^{p-q-2}} = \sum_{n=0}^{\infty} \frac{\Gamma(n+p-q-2)}{n! \, \Gamma(p-q-2)} \, z^n, \qquad (z \in \mathbb{D}),$$

and that

$$\frac{\Gamma(n+p-q-2)}{n! \, \Gamma(p-q-2)} \asymp \frac{1}{n^{q+3-p}}$$

as $n \longrightarrow \infty$.

The following corollary, which is a direct consequence of Lemma 4.3, should be compared with Corollary 4.2.

Corollary 4.4 *Let $\gamma > -1$, let $0 < \alpha < \infty$, and define*

$$f_\alpha(z) = \frac{1}{(1-z)^\alpha}, \qquad (z \in \mathbb{D}).$$

Then

$$f_\alpha \in \bigcap_{0 < p < \frac{2+\gamma}{\alpha}} A_\gamma^p(\mathbb{D}),$$

but

$$f_\alpha \notin A_\gamma^{\frac{2+\gamma}{\alpha}}(\mathbb{D}).$$

Example 4.5 Fix $0 < p < \infty$ and $q > 2p$, and pick any α such that $2/q \le \alpha < 1/p$. Then the function f_α, exploited in Corollaries 4.2 and 4.4, satisfies

$$f_\alpha \in H^p(\mathbb{D}) \qquad \text{but} \qquad f_\alpha \notin A^q(\mathbb{D}).$$

For a Blaschke factor b_w, we have

$$\int_0^{2\pi} \int_0^1 |b_w'(re^{i\theta})|^p (1-r)^q \, r dr d\theta = \int_0^{2\pi} \int_0^1 \frac{(1-|w|^2)^p}{|1 - \bar{w} \, re^{i\theta}|^{2p}} (1-r)^q \, r dr d\theta,$$

and thus, by Lemma 4.3,

$$\int_0^{2\pi} \int_0^1 |b_w'(re^{i\theta})|^p (1-r)^q \, r dr d\theta \asymp \begin{cases} (1-|w|)^p & \text{if } 0 < 2p < q+2, \\[2mm] (1-|w|)^p \log \dfrac{1}{(1-|w|)} & \text{if } \quad 2p = q+2, \\[2mm] (1-|w|)^{q+2-p} & \text{if } \quad 2p > q+2. \end{cases}$$

In each case, the constants involved just depend on p. More specifically, they do not depend on w. In particular,

$$\|b_w'\|_{BP} \asymp \begin{cases} (1-|w|) & \text{if } 0 < p < \frac{1}{2}, \\[2mm] (1-|w|) \log \dfrac{1}{(1-|w|)} & \text{if } \quad p = \frac{1}{2}, \\[2mm] (1-|w|)^{\frac{1}{p}-1} & \text{if } \frac{1}{2} < p < 1, \end{cases} \tag{4.24}$$

where we recall that

$$\|b_w'\|_{BP} = \int_0^{2\pi} \int_0^1 |b_w'(re^{i\theta})| (1-r)^{\frac{1}{p}-2} \, dr d\theta.$$

4.4 The Angular Derivative

Let
$$z \in S_C(e^{i\theta}) = \{ z \in \mathbb{D} : |z - e^{i\theta}| \le C (1 - |z|) \}$$
and consider the circle $D_z = \{ w : |w - z| = (1 - |z|)/2 \}$. Then, for each $w \in D_z$, we have $1 - |z| \le 2(1 - |w|)$, and thus

$$\begin{aligned}
|w - e^{i\theta}| &\le |w - z| + |z - e^{i\theta}| \\
&\le \frac{1 - |z|}{2} + C (1 - |z|) \qquad\qquad (4.25) \\
&\le (2C + 1) (1 - |w|).
\end{aligned}$$

Therefore,
$$D_z \subset S_{2C+1}(e^{i\theta}). \qquad\qquad (4.26)$$

Theorem 4.6 *Let f be analytic on the open unit disc $\mathbb{D}$, and let $e^{i\theta} \in \mathbb{T}$. Then the following two statements are equivalent.*

(i) Both nontangential limits

$$f(e^{i\theta}) = \lim_{\substack{z \to e^{i\theta} \\ \vartriangleleft}} f(z) \qquad and \qquad L = \lim_{\substack{z \to e^{i\theta} \\ \vartriangleleft}} \frac{f(z) - f(e^{i\theta})}{z - e^{i\theta}}$$

exist.
(ii) The function f' has a nontangential limit at $e^{i\theta}$, i.e.

$$f'(e^{i\theta}) = \lim_{\substack{z \to e^{i\theta} \\ \vartriangleleft}} f'(z)$$

exists.

Moreover, under the preceding equivalent conditions, we have $f'(e^{i\theta}) = L$.

Proof. $(i) \Longrightarrow (ii)$: Put

$$g(z) = \frac{f(z) - f(e^{i\theta})}{z - e^{i\theta}} - L, \qquad (z \in \mathbb{D}).$$

Fix a Stolz domain $S_C(e^{i\theta})$. Then, given $\varepsilon > 0$, there is $\delta = \delta(\varepsilon, C)$ such that $|g(z)| < \varepsilon$ for all $z \in S_{2C+1}$ with $|z - e^{i\theta}| < \delta$.

Let $z \in S_C(e^{i\theta})$, and let D_z denote the circle of radius $(1 - |z|)/2$ and center z. Hence, by Cauchy's integral formula,

$$\begin{aligned}
f'(z) &= \frac{1}{2\pi i} \int_{D_z} \frac{f(w)}{(w - z)^2} \, dw \\
&= L + \frac{1}{2\pi i} \int_{D_z} \frac{g(w) (w - e^{i\theta})}{(w - z)^2} \, dw.
\end{aligned}$$

If we further assume that $(1 - |z|) < \delta' = 2\delta/(1+2C)$, then, by (4.25), for each $w \in D_z$, $|w - e^{i\theta}| < \delta$, and thus, by (4.26), we have

$$\left| \frac{g(w)(w - e^{i\theta})}{(w - z)^2} \right| \le \frac{2\varepsilon(1+2C)}{1 - |z|}$$

for all $w \in D_z$. Since the length of D_z is $\pi(1 - |z|)$, we obtain the estimate

$$|f'(z) - L| \le \varepsilon(1+2C)$$

for each $z \in S_C(e^{i\theta})$ with $1 - |z| < \delta'$. This means that $f'(z)$ tends to L, as z nontangentially tends to $e^{i\theta}$.

$(ii) \implies (i)$: The assumption implies that the integral $\int_{[0,e^{i\theta}]} f'(w)\,dw$ is well-defined and we put

$$\lambda = f(0) + \int_{[0,e^{i\theta}]} f'(w)\,dw.$$

Then, by Cauchy's theorem, we have

$$\lambda = f(z) + \int_{[z,e^{i\theta}]} f'(w)\,dw$$

for each $z \in \mathbb{D}$. On the Stolz domain $S_C(e^{i\theta})$, we have

$$\left| \int_{[z,e^{i\theta}]} f'(w)\,dw \right| \le M\,|z - e^{i\theta}|, \qquad \text{where} \qquad M = \sup_{w \in S_C(e^{i\theta})} |f'(w)|.$$

Hence, λ is in fact the nontangential limit of $f(z)$ at the boundary point $e^{i\theta}$. As usual, write $f(e^{i\theta})$ for λ. Then we can say

$$\frac{f(z) - f(e^{i\theta})}{z - e^{i\theta}} = -\frac{1}{z - e^{i\theta}} \int_{[z,e^{i\theta}]} f'(w)\,dw$$

$$= f'(e^{i\theta}) + \frac{1}{z - e^{i\theta}} \int_{[z,e^{i\theta}]} (f'(e^{i\theta}) - f'(w))\,dw.$$

In each Stolz domain, the last integral tends to zero az $z \longrightarrow e^{i\theta}$. This means that

$$\lim_{\substack{z \to e^{i\theta} \\ \vartriangleleft}} \frac{f(z) - f(e^{i\theta})}{z - e^{i\theta}} = f'(e^{i\theta}).$$

4.5 The Carathéodory Derivative

Theorem 4.6 deals with the angular derivative of analytic functions on the open unit disc $\mathbb{D}$, with no restriction on the range of such functions. In this section, we consider the smaller class of analytic functions $f : \mathbb{D} \longrightarrow \mathbb{D}$, i.e. the self-maps of the open unit disc $\mathbb{D}$. We say f has an *angular derivative* in the sense of *Carathéodory* at $e^{i\theta} \in \mathbb{T}$ if

(i) f has an angular derivative at $e^{i\theta}$,
(ii) And, moreover, $|f(e^{i\theta})| = 1$.

Theorem 4.8 gives a complete characterization of functions which have an angular derivative in the sense of Carathéodory. But, first we need an inequality which is interesting in its own right.

Lemma 4.7 (Julia [29]) *Let $f : \mathbb{D} \longrightarrow \mathbb{D}$ be analytic. Suppose that there is a sequence $(z_n)_{n \geq 1}$ in $\mathbb{D}$ such that*

$$\lim_{n \to \infty} z_n = \alpha \in \mathbb{T} \qquad and \qquad \lim_{n \to \infty} f(z_n) = \beta \in \mathbb{T}$$

and

$$\lim_{n \to \infty} \frac{1 - |f(z_n)|}{1 - |z_n|} = A < \infty.$$

Then

$$\frac{|\beta - f(z)|^2}{1 - |f(z)|^2} \leq A \frac{|\alpha - z|^2}{1 - |z|^2}, \qquad (z \in \mathbb{D}).$$

Proof. According to the Schwarz–Pick theorem, we have

$$\left| \frac{f(z) - f(z_n)}{1 - \overline{f(z_n)} f(z)} \right| \leq \left| \frac{z - z_n}{1 - \overline{z}_n z} \right|.$$

It is easy to directly verify that

$$1 - \left| \frac{a - b}{1 - \overline{b} a} \right|^2 = \frac{(1 - |a|^2)(1 - |b|^2)}{|1 - \overline{b} a|^2}, \qquad (a, b \in \mathbb{D}). \tag{4.27}$$

Hence, applying this identity to the preceding inequality gives us

$$\frac{(1 - |f(z)|^2)(1 - |f(z_n)|^2)}{|1 - \overline{f(z_n)} f(z)|^2} \geq \frac{(1 - |z|^2)(1 - |z_n|^2)}{|1 - \overline{z}_n z|^2}.$$

Rewrite this inequality as

$$\frac{|1 - \overline{f(z_n)} f(z)|^2}{1 - |f(z)|^2} \leq \frac{1 - |f(z_n)|^2}{1 - |z_n|^2} \times \frac{|1 - \overline{z}_n z|^2}{1 - |z|^2}.$$

Now, let $n \longrightarrow \infty$.

Julia's inequality implies

$$|\beta - f(z)|^2 \leq A \frac{|\alpha - z|}{1 - |z|} \times |\alpha - z|,$$

and this weaker inequality shows that f has the nontangential limit β at the boundary point α.

Theorem 4.8 (Carathéodory [11]) *Let* $f : \mathbb{D} \longrightarrow \mathbb{D}$ *be analytic, let* $e^{i\theta} \in \mathbb{T}$, *and put*

$$c = \liminf_{z \to e^{i\theta}} \frac{1 - |f(z)|}{1 - |z|}.$$

Then f *has angular derivative in the sense of Carathéodory at* $e^{i\theta}$ *if and only if* $c < \infty$. *Moreover, in this case,* $c = |f'(e^{i\theta})| > 0$, *and*

$$f'(e^{i\theta}) = e^{-i\theta} f(e^{i\theta}) |f'(e^{i\theta})|.$$

Finally, we have

$$\frac{|f(e^{i\theta}) - f(z)|^2}{1 - |f(z)|^2} \leq |f'(e^{i\theta})| \frac{|e^{i\theta} - z|^2}{1 - |z|^2}, \qquad (z \in \mathbb{D}),$$

and the constant $|f'(e^{i\theta})|$ *is the best possible.*

Proof. If f has angular derivative in the sense of Carathéodory at $e^{i\theta}$, then the inequality

$$\frac{1 - |f(re^{i\theta})|}{1 - r} \leq \left| \frac{f(e^{i\theta}) - f(re^{i\theta})}{e^{i\theta} - re^{i\theta}} \right|$$

and the formula

$$f'(e^{i\theta}) = \lim_{\substack{z \to e^{i\theta} \\ \vartriangleleft}} \frac{f(z) - f(e^{i\theta})}{z - e^{i\theta}}$$

from Theorem 4.6 together show that

$$c = \liminf_{z \to e^{i\theta}} \frac{1 - |f(z)|}{1 - |z|} \leq \lim_{r \to 1} \left| \frac{f(re^{i\theta}) - f(e^{i\theta})}{re^{i\theta} - e^{i\theta}} \right| = |f'(e^{i\theta})| < \infty.$$

The other direction is a bit more delicate. Hence, assume that $c < \infty$. This means that there is a sequence $(z_n)_{n \geq 1}$ in $\mathbb{D}$ which converges to $e^{i\theta}$ and $f(z_n)$ converges to a unimodular number, say $\beta \in \mathbb{T}$, such that

$$\frac{1 - |f(z_n)|}{1 - |z_n|} \longrightarrow c.$$

Hence, by Julia's inequality, $c > 0$, $\beta = f(e^{i\theta})$, and, on the radius $[0, e^{i\theta})$, the Julia's inequality implies

$$\left(\frac{1 - |f(re^{i\theta})|}{1 - r}\right)^2 \le \left|\frac{f(e^{i\theta}) - f(re^{i\theta})}{1 - r}\right|^2 \le c\,\frac{1 - |f(re^{i\theta})|^2}{1 - r^2}. \qquad (4.28)$$

Hence, we have

$$\frac{1 - |f(re^{i\theta})|}{1 - r} \le c\,\frac{1 + |f(re^{i\theta})|}{1 + r},$$

and this estimation implies

$$c \le \liminf_{r \to 1} \frac{1 - |f(re^{i\theta})|}{1 - r} \le \limsup_{r \to 1} \frac{1 - |f(re^{i\theta})|}{1 - r} \le c,$$

and thus

$$\lim_{r \to 1} \frac{1 - |f(re^{i\theta})|}{1 - r} = c. \qquad (4.29)$$

In the light of (4.28), we also have

$$\lim_{r \to 1} \left|\frac{f(e^{i\theta}) - f(re^{i\theta})}{1 - r}\right| = \lim_{r \to 1} \left|\frac{1 - \overline{f(e^{i\theta})}\, f(re^{i\theta})}{1 - r}\right| = c. \qquad (4.30)$$

Put $w = \overline{f(e^{i\theta})}\, f(re^{i\theta})$. By (4.29) and (4.30) and that $0 < c < \infty$, we see that $|1 - w|/(1 - |w|) \longrightarrow 1$. This ensures that $\arg w = o(1 - |w|)$, and thus we also have

$$\lim_{r \to 1} \frac{1 - \overline{f(e^{i\theta})}\, f(re^{i\theta})}{1 - r} = c. \qquad (4.31)$$

By Julia's inequality, the function

$$g(z) = \frac{1 - \overline{f(e^{i\theta})}\, f(z)}{1 - e^{-i\theta} z}$$

is bounded on each fixed Stolz domain. Hence, (4.31) along with Lindelöf's theorem imply that

$$c = \lim_{z \xrightarrow{\vartriangleleft} e^{i\theta}} \frac{1 - \overline{f(e^{i\theta})} f(z)}{1 - e^{-i\theta} z}.$$

We can rewrite this formula as

$$c = \lim_{z \xrightarrow{\vartriangleleft} e^{i\theta}} \frac{f(e^{i\theta}) - f(z)}{e^{i\theta} - z} = c\, e^{-i\theta}\, f(e^{i\theta}).$$

Therefore, by Theorem 4.6, the angular derivative in the sense of Carathéodory exists and

$$f'(e^{i\theta}) = c\, e^{-i\theta}\, f(e^{i\theta}).$$

This last inequality immediately implies $c = |f'(e^{i\theta})|$. Finally, to show that the constant c is sharp, consider the combination

$$\frac{|f(e^{i\theta}) - f(z)|^2}{|z - e^{i\theta}|^2} \times \frac{1 - |z|^2}{1 - |f(z)|^2}.$$

Julia's inequality says that this expression is bounded above by c for all $z \in \mathbb{D}$. Moreover, as $z = re^{i\theta} \longrightarrow e^{i\theta}$, (4.29) and (4.30) reveal that the combinations tends to $c^2/c = c$. Hence, the constant c is sharp.

If a function f in the closed unit ball of $H^\infty(\mathbb{D})$ can be written as $f = gh$, where both g and h are also nonconstant functions in the closed unit ball of $H^\infty(\mathbb{D})$, we say that g and h are the *divisors* of f.

Corollary 4.9 *Let f be an element of the closed unit ball of $H^\infty(\mathbb{D})$. Suppose that f has angular derivative in the sense of Carathéodory at $e^{i\theta} \in \mathbb{T}$. Then every divisor of f also has angular derivative in the sense of Carathéodory at $e^{i\theta}$.*

Proof. Let g be a divisor of f. Since $|f| \leq |g|$, we have

$$\frac{1 - |g(re^{i\theta})|}{1 - r} \leq \frac{1 - |f(re^{i\theta})|}{1 - r},$$

which, using Theorem 4.8, implies

$$\liminf_{r \to 1} \frac{1 - |g(re^{i\theta})|}{1 - r} < \infty.$$

The same theorem again ensures that g has angular derivative in the sense of Carathéodory at $e^{i\theta}$.

Based on the value of

$$c = \liminf_{z \longrightarrow e^{i\theta}} \frac{1 - |f(z)|}{1 - |z|},$$

where f is an element of the closed unit ball of $H^\infty(\mathbb{D})$, we can partition this ball into three parts. Julia's inequality (Lemma 4.7) shows that $c = 0$ if and only if f is a unimodular constant. Theorem 4.8 ensures that $0 < c < \infty$ if and only if f is not identically a unimodular constant and has angular derivative in the sense of Carathéodory at $e^{i\theta}$. The third part, corresponding to $c = \infty$, precisely consists of functions which do not have angular derivative in the sense of Cathéodory. In this case, since

$$\lim_{z \longrightarrow e^{i\theta}} \frac{1 - |f(z)|}{1 - |z|} = \infty,$$

we have

$$\lim_{z \underset{\vartriangleleft}{\longrightarrow} e^{i\theta}} \frac{f(e^{i\theta}) - f(z)}{e^{i\theta} - z} = \infty.$$

Convention: Whenever f fails to have derivative in the sense of Carathéodory at $e^{i\theta}$, we write $|f'(e^{i\theta})| = \infty$.

We should be aware that writing $|f'(e^{i\theta})| = \infty$ does not mean or imply that $\lim_{r \to 1} |f'(re^{i\theta})| = \infty$. For example, consider the function $f(z) = z/2$ for which $|f'(e^{i\theta})| = \infty$ at all $e^{i\theta} \in \mathbb{T}$. However, accepting the convention, in any case, we have

$$|f'(e^{i\theta})| = \liminf_{z \to e^{i\theta}} \frac{1 - |f(z)|}{1 - |z|} = \lim_{r \to 1} \frac{1 - |f(re^{i\theta})|}{1 - r}.$$

This convention simplifies the statement of some results. As an example, we invite you to rewrite the following corollaries if we had not accepted the above convention.

Corollary 4.10 *Let f and f_n, $n \geq 1$, be elements of the closed unit ball of $H^\infty(\mathbb{D})$. Suppose that f_n converges uniformly to f on each compact subset of $\mathbb{D}$. Then*

$$|f'(e^{i\theta})| \leq \liminf_{n \to \infty} |f_n'(e^{i\theta})|, \qquad (e^{i\theta} \in \mathbb{D}).$$

Proof. The only interesting case is when

$$M = \liminf_{n \to \infty} |f_n'(e^{i\theta})| < \infty.$$

By Theorem 4.8, we have

$$\frac{|f_n(e^{i\theta}) - f_n(z)|^2}{1 - |f_n(z)|^2} \leq |f_n'(e^{i\theta})| \frac{|e^{i\theta} - z|^2}{1 - |z|^2}, \qquad (z \in \mathbb{D}),$$

for any n such that $|f_n'(e^{i\theta})| < \infty$. Choose a subsequence $(n_k)_{k \geq 1}$ such that

$$\lim_{k \to \infty} |f_{n_k}'(e^{i\theta})| = M$$

and that $f_{n_k}(e^{i\theta})$ tends to a unimodular constat, say $\beta \in \mathbb{T}$. Hence, passing to the limit, we obtain

$$\frac{|\beta - f(z)|^2}{1 - |f(z)|^2} \leq M \frac{|e^{i\theta} - z|^2}{1 - |z|^2}, \qquad (z \in \mathbb{D}).$$

Thus, β is in fact the nontangential limit of f at $e^{i\theta}$. In the first place, the above inequality implies

$$1 - |f(re^{i\theta})| \leq 2M(1 - r).$$

and thus, by Theorem 4.8, $f'(e^{i\theta})$ exists. Secondly, the last assertion of Theorem 4.8 says that the constant $|f'(e^{i\theta})|$ in Julia's inequality is the best possible. This implies that $|f'(e^{i\theta})| \leq M$.

Corollary 4.11 *Let g and h be elements of the closed unit ball of $H^\infty(\mathbb{D})$, and let $f = gh$. Then*

$$|f'(e^{i\theta})| = |g'(e^{i\theta})| + |h'(e^{i\theta})|, \qquad (e^{i\theta} \in \mathbb{T}).$$

Proof. If g and h have angular derivatives in the sense of Carathéodory, then so does f and by elementary calculus

$$f'(e^{i\theta}) = g'(e^{i\theta})\, h(e^{i\theta}) + g(e^{i\theta})\, h'(e^{i\theta}). \qquad (4.32)$$

Then, by the formula in Theorem 4.8, we have

$$f'(e^{i\theta}) = e^{i\theta}\, f(e^{i\theta})\, |f'(e^{i\theta})|,$$

and a similar formula holds for g and h. Replace these three formulas in (4.32) and simply using the identity $f(e^{i\theta}) = g(e^{i\theta})\, h(e^{i\theta})$.

If one of the functions g or h fails to have the angular derivatives in the sense of Carathéodory, then, by Corollary 4.9, so does f and in this case both sides of the identity are ∞.

Corollary 4.12 *Let f and f_n, $n \geq 1$, be elements of the closed unit ball of $H^\infty(\mathbb{D})$. Suppose that each f_n is a divisor of f and moreover f_n converges uniformly to f on compact subsets of $\mathbb{D}$. Then*

$$|f'(e^{i\theta})| = \lim_{n\to\infty} |f_n'(e^{i\theta})|, \qquad (e^{i\theta} \in \mathbb{T}).$$

Proof. Since f_n is a divisor of f, Corollary 4.11 says

$$|f_n'(e^{i\theta})| \leq |f'(e^{i\theta})|, \qquad (e^{i\theta} \in \mathbb{T}, \, n \geq 1).$$

Moreover, by Corollary 4.10,

$$|f'(e^{i\theta})| \leq \liminf_{n\to\infty} |f_n'(e^{i\theta})|, \qquad (e^{i\theta} \in \mathbb{T}).$$

Hence, the result follows.

The following result is a variation of Corollary 4.12.

Corollary 4.13 *Let f_n, $n \geq 1$, be elements of the closed unit ball of $H^\infty(\mathbb{D})$ such that the product*

$$f = \prod_{n=1}^{\infty} f_n$$

converges uniformly on compact subsets of $\mathbb{D}$. Then

$$|f'(e^{i\theta})| = \sum_{n=1}^{\infty} |f_n'(e^{i\theta})|, \qquad (e^{i\theta} \in \mathbb{T}).$$

Proof. Put

$$g_N = \prod_{n=1}^{N} f_n, \qquad (N \geq 1).$$

Then g_N is a divisor of f and g_N converges uniformly to f on compact subsets of $\mathbb{D}$. Moreover, by Corollary 4.11,

$$|g_N'(e^{i\theta})| = \sum_{n=1}^{N} |f_n'(e^{i\theta})|, \qquad (e^{i\theta} \in \mathbb{T}).$$

Now, apply Corollary 4.12.

Corollary 4.14 *Let f_n, $n \geq 1$, be elements of the closed unit ball of $H^\infty(\mathbb{D})$ such that the product*

$$f = \prod_{n=1}^{\infty} f_n$$

converges uniformly on compact subsets of $\mathbb{D}$. Let $0 < p \leq 1$. Then

$$\sum_{n=1}^{\infty} \|f_n'\|_p \leq \|f'\|_p \leq \left(\sum_{n=1}^{\infty} \|f_n'\|_p^p \right)^{1/p}.$$

Remark. The notation $\|\cdot\|_p$ represents the norm in $L^p(\mathbb{T})$. In particular, if the derivative fails to exist on a set of positive measure, then the norm must be replaced by ∞. Hence, in particular, the result will be mainly applied when the functions involved are inner functions and have Carathéodory derivatives almost everywhere on $\mathbb{T}$. Moreover, in this case, we will see that if the L^p-norm is finite, then the derivative is indeed in $H^p(\mathbb{D})$.

Proof. The left inequality is a consequence of Minkowski's inequality and the identity in Corollary 4.13. The right inequality comes from

$$|f'(e^{i\theta})|^p \leq \sum_{n=1}^{\infty} |f_n'(e^{i\theta})|^p, \qquad (e^{i\theta} \in \mathbb{T}).$$

4.6 Another Characterization of the Carathéodory Derivative

In Sect. 4.5, we defined the Carathéodory derivative at a boundary point and gave a complete characterization in Theorem 4.8. This characterization just depends on the absolute value of the function. In this section, we exploit the canonical factorization theorem to give another criterion for the existence of the Carathéodory derivative.

According to the canonical factorization theorem, a function f in the closed unit ball of $H^\infty(\mathbb{D})$ can be written as

$$f(z) = \gamma \prod_{n=1}^{\infty} \frac{|z_n|}{z_n} \frac{z_n - z}{1 - \bar{z}_n z} \times \exp\left(-\int_{\mathbb{T}} \frac{e^{it} + z}{e^{it} - z} \, d\mu(e^{it})\right), \qquad (4.33)$$

where $(z_n)_{n \geq 1}$ is the sequence of zeros of f in $\mathbb{D}$, and μ is a positive Borel measure. More explicitly, μ has the Lebesgue decomposition

$$d\mu(e^{it}) = -\log|f(e^{it})| \frac{dt}{2\pi} + d\sigma(e^{it}),$$

where σ is a finite positive singular Borel measure on $\mathbb{T}$. At this point, we have developed enough tools to give a formula for $|f'(e^{i\theta})|$, the absolute value of the derivative in the sense of Carathéodory at the boundary point $e^{i\theta}$, in terms of z_n and μ.

The following formula for $|f'(e^{i\theta})|$ was first proved by M. Riesz for singular inner functions [41]. Then Frostman proved it for infinite Blaschke products [23]. The general case was proved by Ahern–Clark [2, 3].

Theorem 4.15 *Let f be in the closed unit ball of $H^\infty(\mathbb{D})$ with the decomposition (4.33). Then*

$$|f'(e^{i\theta})| = \sum_{n=1}^{\infty} \frac{1 - |z_n|^2}{|e^{i\theta} - z_n|^2} + 2 \int_{\mathbb{T}} \frac{d\mu(e^{it})}{|e^{i\theta} - e^{it}|^2}$$

for each $e^{i\theta} \in \mathbb{T}$.

Proof. If the measure μ has a positive mass (a Dirac measure) at the point $e^{i\theta}$, then both sides are infinite. As a matter of fact, on the right side, the integral explodes, and for the left side,

$$\lim_{r \to 1} S_\sigma(re^{i\theta}) = 0,$$

where S_σ is the singular inner function constructed with σ. Since $|f| \leq |S_\sigma|$, we also have

$$\lim_{r \to 1} f(re^{i\theta}) = 0,$$

and thus, by definition, the derivative in the sense of Carathéodory at $e^{i\theta}$ does not exist. Hence, by convention, we have $|f'(e^{i\theta})| = \infty$.

Now, suppose that μ has no point mass at $e^{i\theta}$. Let

$$f_m(z) = \gamma \prod_{n=1}^{m} \frac{|z_n|}{z_n} \frac{z_n - z}{1 - \bar{z}_n z} \times \exp\left(-\int_{\mathbb{T}} \frac{e^{it} + z}{e^{it} - z} \, d\mu_m(e^{it})\right),$$

where μ_m is the restriction of μ to the arc

$$\mathbb{T} \setminus \{e^{it} : \theta - \frac{1}{m} < t < \theta + \frac{1}{m}\}.$$

Hence, f_m is a divisor of f and, as $m \longrightarrow \infty$, f_m converges uniformly to f on compact subsets of $\mathbb{D}$. Therefore, by Corollary 4.12,

$$|f'(e^{i\theta})| = \lim_{m \to \infty} |f'_m(e^{i\theta})|.$$

The advantage of f_m is that it is analytic in a neighborhood of $e^{i\theta}$. Hence, by (4.15), we have

$$|f'_m(e^{i\theta})| = \sum_{n=1}^{m} \frac{1 - |z_n|^2}{|e^{i\theta} - z_n|^2} + 2 \int_{\mathbb{T}} \frac{d\mu_m(e^{it})}{|e^{i\theta} - e^{it}|^2}.$$

The general formula now follows from the monotone convergence theorem.

The following result is a special case of Theorem 4.15, where under some mild conditions, both sides of the formula for $|f'(e^{i\theta})|$ are finite.

Corollary 4.16 *Let ϕ be an inner function with the canonical decomposition*

$$\phi(z) = \gamma \prod_{n=1}^{\infty} \frac{|z_n|}{z_n} \frac{z_n - z}{1 - \bar{z}_n z} \times \exp\left(-\int_{\mathbb{T}} \frac{e^{it} + z}{e^{it} - z} d\sigma(e^{it})\right).$$

Suppose that $\phi' \in \mathcal{N}$. Then, for almost all $e^{i\theta} \in \mathbb{T}$,

$$\left|\lim_{r \to 1} \phi'(re^{i\theta})\right| = \sum_{n=1}^{\infty} \frac{1 - |z_n|^2}{|e^{i\theta} - z_n|^2} + 2 \int_{\mathbb{T}} \frac{d\sigma(e^{it})}{|e^{i\theta} - e^{it}|^2}.$$

Proof. By the definition of an inner function,

$$\phi(e^{i\theta}) = \lim_{r \to 1} \phi(re^{i\theta})$$

exists and

$$|\phi(e^{i\theta})| = 1$$

for almost all $e^{i\theta} \in \mathbb{T}$. The assumption $\phi' \in \mathcal{N}$ ensures that

$$\phi'(e^{i\theta}) = \lim_{r \to 1} \phi'(re^{i\theta})$$

also exists for almost all $e^{i\theta} \in \mathbb{T}$. Hence, ϕ has derivative in the sense of Carathéodory almost everywhere on $\mathbb{T}$. At such points, where both radial limits ϕ and ϕ' exist, the absolute value of the derivative is finite and given by the formula in Theorem 4.15.

Chapter 5
H^p-Means of S'

5.1 The Effect of Singular Factors

Let ϕ be an inner function, and let $\phi = BS$ be its canonical decomposition. Then the identity

$$\phi' = B'S + BS'$$

does not give any indication that S might be a divisor of ϕ'. However, this is indeed the case when ϕ' is in the Nevanlinna class. Given an analytic function f on $\mathbb{D}$ whose trace on $\mathbb{T}$ is well-defined, the assumption $f \in L^p(\mathbb{T})$ alone does not imply $f \in H^p(\mathbb{D})$. An extra condition is usually needed. The following result says that for an *inner* function ϕ, $|\phi'| \in L^p(\mathbb{T})$ implies $\phi' \in H^p(\mathbb{D})$.

Theorem 5.1 (Ahern–Clark [2]) *Let ϕ be an inner function with the canonical decomposition $\phi = BS$. Let $0 < p \le \infty$. Then the following are equivalent:*

(i) $\phi'/S \in H^p(\mathbb{D})$;
(ii) $\phi' \in H^p(\mathbb{D})$;
(iii) $|\phi'| \in L^p(\mathbb{T})$.

Remark. In parts (i) and (ii), ϕ'/S and ϕ' are well-defined analytic functions on $\mathbb{D}$, while in part (iii), the notation $|\phi'|$ stands for the absolute value of the Carathéodory derivative of ϕ. We recall that whenever the Carathéoodory derivative does not exist, by convention, we put $|\phi'| = \infty$. See Sect. 4.5.

Proof. The implications $(i) \implies (ii) \implies (iii)$ are trivial. However, we emphasize that, if $\phi' \in H^p(\mathbb{D})$, then the trace of ϕ' at almost all point of $\mathbb{T}$ is in fact the derivative of ϕ in the sense of Carathéodory.

To prove $(iii) \implies (i)$, write $f = \phi'/S$ and $g = S'/S$. Hence, the relation $\phi' = B'S + BS'$ is rewritten as $f = B' + Bg$. Clearly, f and g are analytic on $\mathbb{D}$. Therefore, by (4.16) and (4.17),

$$|f(re^{i\theta})| \le \sum_{n=1}^{\infty} \frac{1 - |z_n|^2}{|1 - \bar{z}_n \, re^{i\theta}|^2} + 2 \int_{\mathbb{T}} \frac{d\sigma(e^{it})}{|re^{i\theta} - e^{it}|^2}$$

for each $re^{i\theta} \in \mathbb{D}$. Since, for each $z \in \overline{\mathbb{D}}$, we have $|1 - rz| \ge |1 - z|/2$, the above estimation implies

$$|f(re^{i\theta})| \le 4 \sum_{n=1}^{\infty} \frac{1 - |z_n|^2}{|e^{i\theta} - z_n|^2} + 8 \int_{\mathbb{T}} \frac{d\sigma(e^{it})}{|e^{i\theta} - e^{it}|^2}.$$

But, by Theorem 4.15, the quantity at the right side of the preceding identity is precisely $4|\phi'(e^{i\theta})|$, i.e.

$$|f(re^{i\theta})| \le 4|\phi'(e^{i\theta})|, \qquad (e^{i\theta} \in \mathbb{T}).$$

Therefore, for all $0 \le r < 1$,

$$\int_0^{2\pi} |f(re^{i\theta})|^p \, d\theta \le 4 \int_0^{2\pi} |\phi'(e^{i\theta})|^p \, d\theta < \infty.$$

This ensures that $f \in H^p(\mathbb{D})$.

A small modification of the proof of Theorem 5.1 yields a similar result when ϕ' is in the Nevanlinna class $\mathcal{N}$.

Theorem 5.2 (Ahern–Clark [2]) *Let ϕ be an inner function with the canonical decomposition $\phi = BS$. Then the following are equivalent:*

(i) $\phi'/S \in \mathcal{N}^+$;
(ii) $\phi' \in \mathcal{N}^+$;
(iii) $\phi' \in \mathcal{N}$;
(iv) $\log^+ |\phi'| \in L^1(\mathbb{T})$.

Remark. See the comment after Theorem 5.1 about $|\phi'|$.

Proof. The implications $(i) \implies (ii) \implies (iii) \implies (iv)$ are trivial. We just need to verify $(iv) \implies (i)$. Assuming, (iv), as in the proof of Theorem 5.1, we have $|f(re^{i\theta})| \le 4|\phi'(e^{i\theta})|$. Thus,

$$\log^+ |f(re^{i\theta})| \le \log^+ |\phi'(e^{i\theta})| + \log 4.$$

This inequality implies two facts. First,

$$\sup_{0 \le r < 1} \int_0^{2\pi} \log^+ |f(re^{i\theta})| \, d\theta < \infty,$$

and second, by the dominated convergence theorem,

$$\lim_{r \to 1} \int_0^{2\pi} \log^+ |f(re^{i\theta})| \, d\theta = \int_0^{2\pi} \log^+ |f(e^{i\theta})| \, d\theta.$$

Therefore, by Theorem 1.18, $f \in \mathcal{N}^+$.

Corollary 5.3 *Let ϕ be an inner function with the canonical decomposition $\phi = BS$. Suppose that $\phi' \in \mathcal{N}$. Then $\phi' \in \mathcal{N}^+$ and S is a divisor of ϕ' in $\mathcal{N}^+$.*

5.2 A Characterization of $\phi' \in H^p(\mathbb{D})$

A variation of the following result has been proved earlier. However, for further reference it is restated below.

Theorem 5.4 *Let ϕ be an inner function with the canonical decomposition*

$$\phi(z) = \gamma \prod_{n=1}^{\infty} \frac{|z_n|}{z_n} \frac{z_n - z}{1 - \bar{z}_n z} \times \exp\left(-\int_{\mathbb{T}} \frac{e^{it} + z}{e^{it} - z} d\sigma(e^{it})\right).$$

Put

$$f_\phi(e^{i\theta}) = \sum_{n=1}^{\infty} \frac{1 - |z_n|^2}{|e^{i\theta} - z_n|^2} + \int_{\mathbb{T}} \frac{d\sigma(e^{it})}{|e^{i\theta} - e^{it}|^2}.$$

Then $\phi' \in H^p(\mathbb{D})$, $0 < p \leq \infty$, if and only if $f_\phi \in L^p(\mathbb{T})$.

Proof. By Theorem 4.15,

$$|\phi'(e^{i\theta})| = \sum_{n=1}^{\infty} \frac{1 - |z_n|^2}{|e^{i\theta} - z_n|^2} + 2\int_{\mathbb{T}} \frac{d\sigma(e^{it})}{|e^{i\theta} - e^{it}|^2}, \qquad (e^{i\theta} \in \mathbb{T}),$$

and clearly $|\phi'| \asymp f_\phi$. Now, apply Theorem 5.1.

Theorem 5.1 and Corollary 4.14 combined together give a strong result about the relation between the derivative of an inner function and the derivative of its divisors. Before stating the theorem, we make a remark. If $(\varphi_n)_{n \geq 1}$ is a sequence of inner functions which converges uniformly on compact subsets of $\mathbb{D}$ to a function ϕ, we cannot deduce that ϕ is inner. As a matter of fact, Carathéodory showed that for each f in the closed unit ball of H^∞, there is a sequence of finite Blaschke products which converges uniformly to f on each compact subset of $\mathbb{D}$. However, if the sequence $(\varphi_n)_{n \geq 1}$ is such that φ_n is a divisor of φ_{n+1} in the family of inner functions, then ϕ is also inner. In this case, it is easier to write

$$\varphi_N = \prod_{n=1}^{N} \phi_n \quad \text{and} \quad \phi = \prod_{n=1}^{\infty} \phi_n.$$

Theorem 5.5 *Let ϕ_n, $n \geq 1$, be inner functions such that the product*

$$\phi = \prod_{n=1}^{\infty} \phi_n$$

converges uniformly on compact subsets of $\mathbb{D}$. Let $0 < p \leq 1$. Then the following assertions hold.

(i) If $\phi' \in H^p(\mathbb{D})$, then $\phi'_n \in H^p(\mathbb{D})$, for all $n \geq 1$, and we have

$$\sum_{n=1}^{\infty} \|\phi'_n\|_p \leq \|\phi'\|_p.$$

(ii) If $\phi'_n \in H^p(\mathbb{D})$, for all $n \geq 1$, with

$$\sum_{n=1}^{\infty} \|\phi'_n\|_p^p < \infty,$$

then $\phi' \in H^p(\mathbb{D})$, and we have

$$\|\phi'\|_p^p \leq \sum_{n=1}^{\infty} \|\phi'_n\|_p^p.$$

In particular, $\phi' \in H^1(\mathbb{D})$ if and only if $\phi'_n \in H^1(\mathbb{D})$, for all $n \geq 1$, and

$$\sum_{n=1}^{\infty} \|\phi'_n\|_1 < \infty.$$

In this situation, we have

$$\|\phi'\|_1 = \sum_{n=1}^{\infty} \|\phi'_n\|_1.$$

5.3 We Never Have $S' \in H^{\frac{1}{2}}(\mathbb{D})$

J. Caughran and A. Shields [13] asked if there is a singular inner function S such that $S' \in H^{1/2}(\mathbb{D})$. M. Cullen [17] constructed a singular inner function S satisfying $S' \in H^p(\mathbb{D})$, for all $0 < p < 1/2$. See Sect. 5.5. Hence, there was a conjecture that there is no singular inner function S with $S' \in H^{1/2}(\mathbb{D})$. The conjecture was proved by Ahern and Clark.

Theorem 5.6 (Ahern–Clark [2]) *Let ϕ be an inner function such that $\phi' \in H^{1/2}(\mathbb{D})$. Then ϕ is a Blaschke product.*

Proof. The inner function ϕ has the canonical decomposition

$$\phi(z) = \gamma \prod_{n=1}^{\infty} \frac{|z_n|}{z_n} \frac{z_n - z}{1 - \bar{z}_n z} \times \exp\left(-\int_{\mathbb{T}} \frac{e^{it} + z}{e^{it} - z} d\sigma(e^{it})\right).$$

Since we assumed $\phi' \in H^{1/2}(\mathbb{D})$, by Corollary 4.16,

$$\phi'(e^{i\theta}) = \lim_{r \to 1} \phi'(re^{i\theta})$$

exists for almost all $e^{i\theta} \in \mathbb{T}$, and its absolute value is given by

$$|\phi'(e^{i\theta})| = \sum_{n=1}^{\infty} \frac{1 - |z_n|^2}{|e^{i\theta} - z_n|^2} + 2 \int_{\mathbb{T}} \frac{d\sigma(e^{it})}{|e^{i\theta} - e^{it}|^2}.$$

In particular, we have

$$|\phi'(e^{i\theta})| \geq \int_{\mathbb{T}} \frac{d\sigma(e^{it})}{|e^{i\theta} - e^{it}|^2}, \qquad (a.e.\ e^{i\theta} \in \mathbb{T}).$$

By the Cauchy–Schwarz inequality,

$$\left(\int_{\mathbb{T}} \frac{d\sigma(e^{it})}{|e^{i\theta} - e^{it}|}\right)^2 \leq \left(\int_{\mathbb{T}} \frac{d\sigma(e^{it})}{|e^{i\theta} - e^{it}|^2}\right) \times \int_{\mathbb{T}} d\sigma(e^{it}).$$

Therefore, the last two inequalities imply

$$\int_{\mathbb{T}} \frac{d\sigma(e^{it})}{|e^{i\theta} - e^{it}|} \leq \|\sigma\|^{1/2} |\phi'(e^{i\theta})|^{1/2}, \qquad (a.e.\ e^{i\theta} \in \mathbb{T}).$$

Take the integral of both sides with respect to the normalized Lebesgue measure and apply Fubini's theorem to obtain

$$\int_{\mathbb{T}} \left(\int_{\mathbb{T}} \frac{d\theta}{|e^{i\theta} - e^{it}|}\right) d\sigma(e^{it}) \leq 2\pi \|\sigma\|^{1/2} \|\phi'\|_{1/2}^2 < \infty.$$

However, we have

$$\int_{\mathbb{T}} \frac{d\theta}{|e^{i\theta} - e^{it}|} = \infty$$

for *any* $e^{it} \in \mathbb{T}$. Therefore, we must have $\sigma = 0$. In other words, ϕ is a Blaschke product.

Corollary 5.7 *Let S be any singular inner function for $\mathbb{D}$. Then*

$$\frac{S'}{S} \notin H^{\frac{1}{2}}(\mathbb{D}).$$

Proof. By Theorem 5.6, $S' \notin H^{\frac{1}{2}}(\mathbb{D})$. Hence, Theorem 5.1 ensures that in fact $S'/S \notin H^{\frac{1}{2}}(\mathbb{D})$.

5.4 The Distance Function

Let K be a closed subset of $[0, 1]$. Then, for each $x \in [0, 1]$, we define

$$d_K(x) = \text{dist}(x, K) = \inf_{t \in K} |t - x|.$$

When there is no confusion, we simply write $d(x)$ instead of $d_K(x)$. The open set $[0, 1] \setminus K$ is at most a countable union of open intervals of $[0, 1]$, i.e.

$$[0, 1] \setminus K = \bigcup_{n=1}^{\infty} I_n.$$

There might be a finite number of such intervals. In this case, either we replace ∞ by a natural number, or assume that $I_n = \emptyset$ from some index on. These intervals are called the *complementary intervals* of K. The length of I_n is denoted by ℓ_n. Clearly, K has Lebesgue measure zero if and only if

$$\sum_{n=1}^{\infty} \ell_n = 1.$$

Lemma 5.8 *Let K be a closed subset of $[0, 1]$. Suppose that $|K| = 0$. Write*

$$[0, 1] \setminus K = \bigcup_{n=1}^{\infty} I_n,$$

and let ℓ_n, $n \geq 1$, denote the length of I_n. Let $0 < \alpha < 1$. Then

$$\int_0^1 \frac{dx}{d_K^{\alpha}(x)} = \frac{2^{\alpha}}{1 - \alpha} \sum_{n=1}^{\infty} \ell_n^{1-\alpha}.$$

Proof. Since K is of Lebesgue measure zero, then

$$\int_0^1 \frac{dx}{d_K^{\alpha}(x)} = \sum_{n=1}^{\infty} \int_{I_n} \frac{dx}{d_K^{\alpha}(x)}.$$

Write $I_n = [a_n, b_n]$, where $\ell_n = b_n - a_n$. Then we have

$$\int_{I_n} \frac{dx}{d_K^\alpha(x)} = \int_{a_n}^{(a_n+b_n)/2} \frac{dx}{(x-a_n)^\alpha} + \int_{(a_n+b_n)/2}^{b_n} \frac{dx}{(b_n-x)^\alpha}$$

$$= 2 \int_0^{\ell_n/2} \frac{dt}{t^\alpha}$$

$$= \frac{2^\alpha}{1-\alpha} \ell_n^{1-\alpha}.$$

In the same manner, we can consider the closed subsets of the unit circle $\mathbb{T}$ and define the distance function d_K. There is a little doubt in this definition, since, based on our needs, we may define d_K either by

$$d_K(e^{i\theta}) = \operatorname{dist}_{\mathbb{C}}(e^{i\theta}, K) = \inf_{e^{it} \in K} |e^{it} - e^{i\theta}|,$$

or by

$$d_K(e^{i\theta}) = \operatorname{dist}_{\mathbb{T}}(e^{i\theta}, K) = \inf_{e^{it} \in K} |t - \theta|,$$

where we confine $|t - \theta|$ to the range $[0, \pi]$. However,

$$|e^{it} - e^{i\theta}| = 2 \left| \sin\left(\frac{t-\theta}{2}\right) \right| \asymp |t - \theta|,$$

and thus, in the applications, it is not ultimately important which definition we have adopted. The final outcome is that if K is any closed subset of $\mathbb{T}$ with Lebesgue measure zero, then

$$\int_{\mathbb{T}} \frac{d\theta}{d_K^\alpha(e^{i\theta})} \asymp \sum_{n=1}^\infty \ell_n^{1-\alpha}, \qquad (0 < \alpha < 1), \qquad (5.1)$$

where $(\ell_n)_{n\geq 1}$ represent the length of the complementary arcs of K. The constants involved in (5.1) do not depend on K.

Lemma 5.9 *Let $z_n = r_n e^{i\theta_n}$, $n \geq 1$, be a sequence satisfying the condition*

$$A = \sum_{n=1}^\infty (1 - r_n)^\alpha < \infty,$$

for some $0 < \alpha \leq 1$. Let

$$d(e^{i\theta}) = \inf_{n\geq 1} |e^{i\theta} - e^{i\theta_n}|, \qquad (e^{i\theta} \in \mathbb{T}).$$

Then, for each $\beta \geq 1 - \alpha$,

$$\sum_{n=1}^\infty \frac{1 - |z_n|}{|e^{i\theta} - z_n|^\beta} \leq \frac{2^\beta A}{d^{\beta+\alpha-1}(e^{i\theta})}, \qquad (e^{i\theta} \in \mathbb{T}).$$

Proof. To estimate the sum, we decompose it as

$$\sum_{n=1}^{\infty} = \sideset{}{'}\sum + \sideset{}{''}\sum,$$

where $\sideset{}{'}\sum$ extends over all n for which $1 - r_n < |e^{i\theta} - e^{i\theta_n}|$, and $\sideset{}{''}\sum$ extends over the complement. For the first sum, since, for each $z \in \overline{\mathbb{D}}$, we have

$$|1 - rz| \geq |1 - z|/2,$$

then

$$\sideset{}{'}\sum \frac{1 - |z_n|}{|e^{i\theta} - z_n|^{\beta}} \leq 2^{\beta} \sideset{}{'}\sum \frac{1 - r_n}{|e^{i\theta} - e^{i\theta_n}|^{\beta}}$$

$$= 2^{\beta} \sideset{}{'}\sum \left(\frac{1 - r_n}{|e^{i\theta} - e^{i\theta_n}|} \right)^{1-\alpha} \frac{(1 - r_n)^{\alpha}}{|e^{i\theta} - e^{i\theta_n}|^{\beta+\alpha-1}}$$

$$\leq 2^{\beta} \sideset{}{'}\sum \frac{(1 - r_n)^{\alpha}}{|e^{i\theta} - e^{i\theta_n}|^{\beta+\alpha-1}}$$

$$\leq 2^{\beta} \frac{\sideset{}{'}\sum (1 - r_n)^{\alpha}}{d^{\beta+\alpha-1}(e^{i\theta})},$$

and similarly, for the second sum,

$$\sideset{}{''}\sum \frac{1 - |z_n|}{|e^{i\theta} - z_n|^{\beta}} \leq \sideset{}{''}\sum \frac{1}{(1 - r_n)^{\beta-1}}$$

$$\leq \sideset{}{''}\sum \left(\frac{1 - r_n}{|e^{i\theta} - e^{i\theta_n}|} \right)^{\beta+\alpha-1} \frac{1}{(1 - r_n)^{\beta-1}}$$

$$= \sideset{}{''}\sum \frac{(1 - r_n)^{\alpha}}{|e^{i\theta} - e^{i\theta_n}|^{\beta+\alpha-1}}$$

$$\leq \frac{\sideset{}{''}\sum (1 - r_n)^{\alpha}}{d^{\beta+\alpha-1}(e^{i\theta})}.$$

Note that, under the assumptions of Lemma 5.9, if we define

$$K = \text{Clos}_{\mathbb{T}}\{e^{i\theta_n} : n \geq 1\},$$

Then $d(e^{i\theta})$ is precisely $d_K(e^{i\theta})$, as defined in the beginning of this section. In the following corollary, we use a special case of Lemma 5.9.

Corollary 5.10 *Let r_n, $n \geq 1$, be a sequence in $[0,1)$ satisfying the condition*

$$A = \sum_{n=1}^{\infty} (1 - r_n)^{\alpha} < \infty,$$

for some $0 < \alpha \leq 1$. Then, for each $\beta \geq 1 - \alpha$,

$$\sum_{n=1}^{\infty} \frac{1 - r_n}{|e^{i\theta} - r_n|^{\beta}} \leq \frac{A}{|1 - e^{i\theta}|^{\beta + \alpha - 1}}, \qquad (e^{i\theta} \in \mathbb{T}).$$

In particular, for each $\beta > 1 - \alpha$,

$$\sum_{n=1}^{\infty} \frac{1 - r_n}{|e^{i\theta} - r_n|^{\beta}} \in \bigcap_{0 < p < \frac{1}{\beta + \alpha - 1}} L^p(\mathbb{T}).$$

Proof. In this case, the zeros accumulate at the point $1 \in \mathbb{T}$. Thus,

$$d(e^{i\theta}) = |1 - e^{i\theta}|.$$

Hence, by Lemma 5.9, or even by direct verification (which reveals that the factor 2^{β} is not needed), we get the proposed estimation. Since

$$\frac{1}{|1 - e^{i\theta}|} \asymp \frac{1}{|\theta|}, \qquad (-\pi < \theta < \pi),$$

we immediately deduce the second result.

5.5 A Construction of S with $S' \in H^p(\mathbb{D})$ for All $0 < p < \frac{1}{2}$

There are plenty of Blaschke products satisfying the growth restriction $B' \in H^{1/2}(\mathbb{D})$. For example, finite Blaschke products fulfill this property. Moreover, a family of infinite Blaschke products with $B' \in H^{1/2}(\mathbb{D})$ was discovered by D. Protas. See Theorem 8.6.

An immediate consequence of Theorem 5.6 is that if S is a nonconstant singular inner function, then certainly

$$S' \notin H^{1/2}(\mathbb{D}).$$

However, the possibility

$$S' \in H^p(\mathbb{D}),$$

for some $0 < p < 1/2$, is not excluded. The following construction shows that this is indeed possible. We remind the reader that if σ is a singular measure, it is carried on a set of Lebesgue measure zero. However, its support (the smallest closed subset of $\mathbb{T}$ which contains a carrier) could be quite large, e.g. it can be even the whole unit circle $\mathbb{T}$.

Lemma 5.11 *Let σ be a positive singular Borel measure on $\mathbb{T}$ such that its support is of Lebesgue measure zero. Denote the complementary arcs of the support by $(I_n)_{n\geq 1}$, and their length by $(\ell_n)_{n\geq 1}$. Suppose that there is an $\alpha \in (0,1)$ such that*

$$\sum_{n=1}^{\infty} \ell_n^{\alpha} < \infty.$$

Let ψ be any element of $L^1(\sigma)$, let $\beta > 1 - \alpha$, and define

$$f(z) = \int_{\mathbb{T}} \frac{\psi(e^{it})}{(e^{it} - z)^{\beta}} \, d\sigma(e^{it}), \qquad (z \in \mathbb{D}).$$

Then $f \in H^p(\mathbb{D})$, where $p = (1-\alpha)/\beta$.

Proof. Write $z = re^{i\theta}$. Then

$$|e^{it} - re^{i\theta}| \geq \frac{1}{2} |e^{it} - e^{i\theta}|.$$

Let

$$d(e^{i\theta}) = \text{dist}(e^{i\theta}, \text{supp } \sigma) = \inf_{e^{it} \in \text{ supp } \sigma} |e^{it} - e^{i\theta}|.$$

Hence, for each $re^{i\theta} \in \mathbb{D}$, we obtain the estimation

$$|f(re^{i\theta})| \leq 2^{\beta} \int_{\mathbb{T}} \frac{|\psi(e^{it})|}{|e^{it} - e^{i\theta}|^{\beta}} \, d\sigma(e^{it}) \leq 2^q \frac{\|\psi\|_{L^1(\sigma)}}{d^{\beta}(e^{i\theta})}.$$

Therefore, for each $r < 1$,

$$\int_0^{2\pi} |f(re^{i\theta})|^p \, d\theta \lesssim \int_0^{2\pi} \frac{d\theta}{d^{\beta p}(e^{i\theta})}.$$

But, by (5.1),

$$\int_0^{2\pi} \frac{d\theta}{d^{\beta p}(e^{i\theta})} \lesssim \sum_{n=1}^{\infty} \ell_n^{1-\beta p} = \sum_{n=1}^{\infty} \ell_n^{\alpha} < \infty.$$

Theorem 5.12 (Cullen [17]) *Let $0 < p < 1/2$. Let σ be a positive singular Borel measure on $\mathbb{T}$ such that its support is of Lebesgue measure zero and the complementary arcs $(I_n)_{n\geq 1}$ of the support are of length $(\ell_n)_{n\geq 1}$ with*

$$\sum_{n=1}^{\infty} \ell_n^{1-2p} < \infty.$$

Let

$$S(z) = \exp\left(-\int_{\mathbb{T}} \frac{e^{it} + z}{e^{it} - z} \, d\sigma(e^{it})\right), \qquad (z \in \mathbb{D}).$$

Then $S' \in H^p(\mathbb{D})$.

Proof. Since $S' = fS$, where

$$f(z) = -\int_{\mathbb{T}} \frac{2e^{it}}{(e^{it} - z)^2} \, d\sigma(e^{it}), \qquad (z \in \mathbb{D}),$$

and by Lemma 5.11, $f \in H^p(\mathbb{D})$, we have $S' \in H^p(\mathbb{D})$.

The conditions of Theorem 5.12 are clearly fulfilled by a singular measure consisting a finite number of Dirac measures. Hence, we easily obtain the following interesting result.

Corollary 5.13 Let $\sigma_1, \sigma_2, \ldots, \sigma_n > 0$ and let $\zeta_1, \zeta_2, \ldots, \zeta_n \in \mathbb{T}$. Put

$$S(z) = \exp\left(-\sum_{k=1}^{n} \sigma_k \frac{\zeta_k + z}{\zeta_k - z}\right), \qquad (z \in \mathbb{D}).$$

Then

$$S' \in \bigcap_{0 < p < 1/2} H^p(\mathbb{D}),$$

but

$$S' \notin H^{1/2}(\mathbb{D}).$$

Chapter 6
B^p-Means of S'

For an inner function ϕ, the H^p-means of ϕ' are not necessarily finite. We will even provide an example of a Blaschke product B such that $B' \notin \mathcal{N}$. See Theorem 7.12. However, for B^p-means the story is different. In this chapter, we study the B^p-means of a general inner function and pay special attention to singular inner functions. In Chap. 9, we come back to this subject again and focus on B^p-means of Blaschke products.

6.1 We Always Have $\phi' \in \cap_{0<p<\frac{1}{2}} B^p(\mathbb{D})$

The following result is indeed one of the oldest in studying the derivative of inner functions.

Theorem 6.1 (Duren–Romberg–Shields [19]) *Let ϕ be any inner function for $\mathbb{D}$. Then*

$$\phi' \in \bigcap_{0<p<\frac{1}{2}} B^p(\mathbb{D}).$$

Proof. By (4.18),

$$|\phi'(re^{i\theta})| \leq \frac{1}{2(1-r)}, \qquad (re^{i\theta} \in \mathbb{D}).$$

Hence,

$$\int_0^{2\pi} \int_0^1 |\phi'(re^{i\theta})| \, (1-r)^{\frac{1}{p}-2} dr d\theta \leq \pi \int_0^1 (1-r)^{\frac{1}{p}-3} dr,$$

and the last integral is finite provided that $0 < p < \frac{1}{2}$.

The result is sharp in the sense that there is an inner function ϕ such that $\phi' \notin B^{\frac{1}{2}}(\mathbb{D})$. For singular inner functions, we can slightly generalize Theorem 6.1.

Theorem 6.2 (Cullen [17]) *Let S be any singular inner function for $\mathbb{D}$. Then*

$$\frac{S'}{S} \in \bigcap_{0<p<\frac{1}{2}} B^p(\mathbb{D}).$$

Proof. First, note that S'/S is an analytic function on $\mathbb{D}$. Second, by (4.10),

$$\left| \frac{S'(re^{i\theta})}{S(re^{i\theta})} \right| = \int_{\mathbb{T}} \frac{d\sigma(e^{it})}{|e^{it} - re^{i\theta}|^2}, \qquad (re^{i\theta} \in \mathbb{D}).$$

Hence, by Fubini's theorem,

$$\int_0^{2\pi} \left| \frac{S'(re^{i\theta})}{S(re^{i\theta})} \right| d\theta = \int_0^{2\pi} \left(\int_{\mathbb{T}} \frac{d\sigma(e^{it})}{|e^{it} - re^{i\theta}|^2} \right) d\theta$$

$$= \int_{\mathbb{T}} \left(\int_0^{2\pi} \frac{d\theta}{|e^{it} - re^{i\theta}|^2} \right) d\sigma(e^{it})$$

$$= \int_{\mathbb{T}} \left(\int_0^{2\pi} \frac{ds}{|1 - re^{is}|^2} \right) d\sigma(e^{it})$$

$$= \frac{2\pi \|\sigma\|}{1 - r^2}.$$

Finally, as in the proof of Theorem 6.1, we conclude

$$\int_0^{2\pi} \int_0^1 \left| \frac{S'(re^{i\theta})}{S(re^{i\theta})} \right| (1-r)^{\frac{1}{p}-2} dr d\theta \leq 2\pi \|\sigma\| \int_0^1 (1-r)^{\frac{1}{p}-3} dr,$$

and the last integral is finite provided that $0 < p < \frac{1}{2}$.

6.2 When Does $\phi' \in B^p(\mathbb{D})$ for Some $\frac{1}{2} \leq p < 1$?

In Sect. 5.2, we gave a necessary and sufficient condition for $\phi' \in H^p(\mathbb{D})$ in terms of the zeros of ϕ in $\mathbb{D}$ and the corresponding singular measure on $\mathbb{T}$. In this section, we treat a similar question for $\phi' \in B^p(\mathbb{D})$. However, our characterization is mainly based on a growth restriction on $|\phi|$.

Theorem 6.3 (Ahern–Clark [4]) *Fix $\frac{1}{2} < p < 1$. Let ϕ be an inner function for $\mathbb{D}$. Then*

$$\|\phi'\|_{B^p} \leq 2 \int_0^{2\pi} \int_0^1 \left(1 - |\phi(re^{i\theta})|\right)(1-r)^{\frac{1}{p}-3} dr d\theta \leq \frac{2p}{2p-1} \|\phi'\|_{B^p}.$$

Proof. By (4.18),

$$|\phi'(re^{i\theta})| \leq \frac{1 - |\phi(re^{i\theta})|^2}{1-r^2} \leq \frac{2(1 - |\phi(re^{i\theta})|)}{1-r}, \qquad (re^{i\theta} \in \mathbb{D}).$$

This estimation immediately implies the first inequality, i.e.

$$\|\phi'\|_{B^p} \leq 2 \int_0^{2\pi} \int_0^1 \left(1 - |\phi(re^{i\theta})|\right)(1-r)^{\frac{1}{p}-3} dr d\theta.$$

To prove the second inequality, since ϕ is an inner function, we have

$$\int_{[re^{i\theta}, e^{i\theta}]} \phi'(z)\, dz = \phi(e^{i\theta}) - \phi(re^{i\theta}), \qquad (a.e.\ e^{i\theta} \in \mathbb{T}).$$

Hence,

$$1 - |\phi(re^{i\theta})| \leq \int_r^1 |\phi'(te^{i\theta})|\, dt, \qquad (a.e.\ e^{i\theta} \in \mathbb{T}).$$

Therefore,

$$\int_0^1 \left(1 - |\phi(re^{i\theta})|\right)(1-r)^{\frac{1}{p}-3} dr \leq \int_0^1 \left(\int_r^1 |\phi'(te^{i\theta})|\, dt\right)(1-r)^{\frac{1}{p}-3} dr$$

$$= \int_0^1 |\phi'(te^{i\theta})| \left(\int_0^t (1-r)^{\frac{1}{p}-3}\, dr\right) dt$$

$$= \frac{p}{2p-1} \int_0^1 |\phi'(te^{i\theta})| \left((1-t)^{\frac{1}{p}-2} - 1\right) dt$$

$$\leq \frac{p}{2p-1} \int_0^1 |\phi'(te^{i\theta})| (1-t)^{\frac{1}{p}-2}\, dt.$$

Integrating with respect to θ gives the second inequality, i.e.

$$\int_0^{2\pi} \int_0^1 \left(1 - |\phi(re^{i\theta})|\right)(1-r)^{\frac{1}{p}-3} dr d\theta \leq \frac{p}{2p-1} \|\phi'\|_{B^p}.$$

Corollary 6.4 *Fix $\frac{1}{2} < p < 1$. Let ϕ be an inner function for $\mathbb{D}$, and let ψ be a divisor of ϕ. Suppose that $\phi' \in B^p(\mathbb{D})$. Then*

$$\psi' \in B^p(\mathbb{D}).$$

Proof. Since ψ is a divisor of ϕ, then $|\phi| \leq |\psi|$, on $\mathbb{D}$. Hence, by Theorem 6.3,

$$\|\psi'\|_{B^p} \leq 2 \int_0^{2\pi} \int_0^1 \left(1 - |\psi(re^{i\theta})|\right) (1-r)^{\frac{1}{p}-3} dr d\theta$$

$$\leq 2 \int_0^{2\pi} \int_0^1 \left(1 - |\phi(re^{i\theta})|\right) (1-r)^{\frac{1}{p}-3} dr d\theta$$

$$\leq \frac{2p}{2p-1} \|\phi'\|_{B^p}.$$

For the following results, we remind the reader that $f \gtrsim g$ means that there is a positive constant C such that $|f| \geq C |g|$.

Corollary 6.5 *Let ϕ be an inner function for $\mathbb{D}$. Suppose that there is a constant $0 < \delta < 1$ such that*

$$\int_0^{2\pi} \left(1 - |\phi(re^{i\theta})|\right) d\theta \gtrsim (1-r)^{\delta}.$$

Then $\phi' \notin B^{\frac{1}{2-\delta}}(\mathbb{D})$.

Proof. The assumption implies

$$\int_0^{2\pi} \int_0^1 \left(1 - |\phi(re^{i\theta})|\right) (1-r)^{\frac{1}{p}-3} dr d\theta = \infty,$$

where $p = 1/(2 - \delta)$. Thus, by Theorem 6.3, $\phi' \notin B^{\frac{1}{2-\delta}}(\mathbb{D})$.

Corollary 6.6 *Let ϕ be an inner function for $\mathbb{D}$. Suppose that there is a constant $0 < \delta \leq 1$ such that*

$$\int_0^{2\pi} \left(1 - |\phi(re^{i\theta})|\right) d\theta \lesssim (1-r)^{\delta}.$$

Then

$$\phi' \in \bigcap_{0 < p < \frac{1}{2-\delta}} B^p(\mathbb{D}).$$

Proof. Fix $\frac{1}{2} < p < \frac{1}{2-\delta}$. The assumption implies

$$\int_0^{2\pi} \int_0^1 \left(1 - |\phi(re^{i\theta})|\right) (1-r)^{\frac{1}{p}-3} dr d\theta < \infty.$$

Thus, by Theorem 6.3, $\phi' \in B^p(\mathbb{D})$.

6.3 We Never Have $S' \in B^{\frac{2}{3}}(\mathbb{D})$

Parallel to the question which was explored in Sect. 5.3, there was a question about the possibility of $S' \in B^p(\mathbb{D})$. In this regard, on the one hand, H. Allen and C. Belna [5] give examples of singular inner functions S with $S' \in B^p(\mathbb{D})$, for all $p < 2/3$, and, on the other hand, P. Ahern and D. Clark [4] showed that there is no singular inner S such that $S' \in B^{2/3}(\mathbb{D})$. These results are treated in this section and the one after.

Corollary 6.5 provides a sufficient condition to ensure that $\phi' \notin B^p(\mathbb{D})$. Hence, we naturally look for conditions to put on an inner function ϕ, which guarantees that the lower estimation

$$\int_0^{2\pi} \left(1 - |\phi(re^{i\theta})|\right) d\theta \gtrsim (1-r)^{\delta}$$

holds for a certain $0 < \delta < 1$. Two such results are treated below. We remind that

$$S_{C,\delta}(e^{i\theta_0}) = \{z \in \mathbb{D} : |e^{i\theta_0} - z| \le C(1 - |z|)^{\delta}\},$$

where $C \ge 1$ and $0 < \delta \le 1$.

Lemma 6.7 *Let ϕ be an inner function for $\mathbb{D}$. Suppose that there are $e^{i\theta_0} \in \mathbb{T}$, $C \ge 1$, $0 < \delta \le 1$, and $0 < M_0 < 1$, such that*

$$|\phi(z)| \le M_0, \qquad \left(z \in S_{C,\delta}(e^{i\theta_0})\right).$$

Then

$$\int_0^{2\pi} \left(1 - |\phi(re^{i\theta})|\right) d\theta \gtrsim (1-r)^{\delta}.$$

In particular, $\phi' \notin B^{\frac{1}{2-\delta}}(\mathbb{D})$.

Proof. Fix $0 < r < 1$. Let $\theta_r \in (0, \pi)$ be such that

$$|1 - re^{i\theta_r}| = C(1-r)^{\delta}.$$

Since $|1 - re^{i\theta_r}| \asymp (1-r) + \theta_r$, we must have

$$\theta_r \asymp (1-r)^{\delta}.$$

Note that θ_r is defined so that the set identity

$$S_{C,\delta}(e^{i\theta_0}) \cap \{re^{i\theta} : -\pi \le \theta \le \pi\} = \{re^{i\theta} : -\theta_r \le \theta - \theta_0 \le \theta_r\}$$

holds. Since $(1 - |\phi|) \ge 0$, we have

$$\int_0^{2\pi} \left(1 - |\phi(re^{i\theta})|\right) d\theta \geq \int_{|\theta - \theta_0| < \theta_r} \left(1 - |\phi(re^{i\theta})|\right) d\theta$$

$$\geq \int_{|\theta - \theta_0| < \theta_r} (1 - M_0) d\theta$$

$$= 2(1 - M_0) \theta_r \gtrsim (1 - r)^\delta.$$

The conclusion $\phi' \notin B^{\frac{1}{2-\delta}}(\mathbb{D})$ follows from Corollary 6.5.

The following result was found by C. Belna and B. Muckenhoupt for the special case

$$S(z) = \exp\left(-\frac{1+z}{1-z}\right), \qquad (z \in \mathbb{D}).$$

Then it was extended to all singular inner functions by P. Ahern and D. Clark.

Lemma 6.8 (Ahern–Clark [4], Belna–Muckenhoupt [6]) *Let S be any singular inner function for $\mathbb{D}$. Then*

$$\int_0^{2\pi} \left(1 - |S(re^{i\theta})|\right) d\theta \gtrsim (1-r)^{\frac{1}{2}}.$$

In particular $S' \notin B^{2/3}(\mathbb{D})$.

Proof. First suppose that S has a factor of the form

$$S_0(z) = \exp\left(-\sigma_0 \frac{\zeta_0 + z}{\zeta_0 - z}\right), \qquad (z \in \mathbb{D}),$$

where $\sigma_0 > 0$ and $\zeta_0 \in \mathbb{T}$. Then we have

$$|S_0(z)| = \exp\left(-\sigma_0 \frac{1 - |z|^2}{|\zeta_0 - z|^2}\right), \qquad (z \in \mathbb{D}).$$

Therefore, $|S_0(z)| \leq e^{-\sigma_0/2}$ on the set

$$\{z \in \mathbb{D} : |\zeta_0 - z|^2 \leq 2(1 - |z|^2)\}.$$

But, this set, which is a disc in $\mathbb{D}$ tangent to $\mathbb{T}$ at ζ_0, contains the domain $S_{\sqrt{2}, 1/2}(\zeta_0)$. Hence, by Lemma 6.7,

$$\int_0^{2\pi} \left(1 - |S(re^{i\theta})|\right) d\theta \geq \int_0^{2\pi} \left(1 - |S_0(re^{i\theta})|\right) d\theta \gtrsim (1-r)^{\frac{1}{2}}.$$

Now, suppose that

$$S(z) = \exp\left(-\int_{\mathbb{T}} \frac{e^{it} + z}{e^{it} - z} d\sigma(e^{it})\right),$$

where σ is a positive singular measure with no atomic mass. Therefore,

$$\frac{1 - |S(re^{i\theta})|^2}{1 - r^2} = \int_0^{2\pi} \frac{|S_t(re^{i\theta})|^2}{|e^{it} - re^{i\theta}|^2} \, d\sigma(e^{it}),$$

where

$$S_t(re^{i\theta}) = \exp\left(-\int_0^t \frac{e^{i\tau} + z}{e^{i\tau} - z} \, d\sigma(e^{i\tau})\right).$$

This representation implies

$$\int_0^{2\pi} \left(1 - |S(re^{i\theta})|\right) d\theta \geq \frac{1}{2} \int_0^{2\pi} \left(1 - |S(re^{i\theta})|^2\right) d\theta$$

$$= \frac{1 - r^2}{2} \int_0^{2\pi} \left(\int_0^{2\pi} \frac{|S_t(re^{i\theta})|^2}{|e^{it} - re^{i\theta}|^2} \, d\sigma(e^{it})\right) d\theta$$

$$\geq \frac{1 - r^2}{2} \int_0^{2\pi} \left(\int_{t+\theta_r}^{t+\Theta_r} \frac{|S_t(re^{i\theta})|^2}{|e^{it} - re^{i\theta}|^2} \, d\theta\right) d\sigma(e^{it}),$$

where θ_r and Θ_r are small positive angles defined by the equations

$$|1 - e^{i\theta_r}|^2 = (1 - r) \qquad \text{and} \qquad |1 - e^{i\Theta_r}|^2 = 4(1 - r).$$

Hence, if $\theta \in [t + \theta_r, t + \Theta_r]$, we have

$$(1 - r) \leq |e^{it} - e^{i\theta}|^2 \leq 4(1 - r).$$

As a consequence, since

$$|e^{it} - re^{i\theta}|^2 = (1 - r)^2 + r |e^{it} - e^{i\theta}|^2,$$

we have

$$(1 - r) \leq |e^{it} - re^{i\theta}|^2 \leq 4(1 - r)$$

provided that $t + \theta_r \leq \theta \leq t + \Theta_r$. Moreover,

$$|S_t(re^{i\theta})| = \exp\left(-\int_0^t \frac{1 - r^2}{|e^{i\tau} - re^{i\theta}|^2} \, d\sigma(e^{i\tau})\right)$$

$$\geq \exp\left(-\int_0^t \frac{1 - r^2}{1 - r} \, d\sigma(e^{i\tau})\right)$$

$$\geq \exp\left(-2\|\sigma\|\right).$$

Therefore, we obtain

$$\int_0^{2\pi} \left(1 - |\phi(re^{i\theta})|\right) d\theta \geq \frac{1-r^2}{2} \int_0^{2\pi} \left(\int_{t+\Theta_r}^{t+\Theta_r} \frac{\exp(-4\|\sigma\|)}{4(1-r)} d\theta \right) d\sigma(e^{it})$$

$$\geq \frac{\|\sigma\| \exp(-4\|\sigma\|)}{8} \Theta_r.$$

Since $|1 - e^{it}| \sim t$, as $t \longrightarrow 0^+$, the defining equation of Θ_r implies

$$\Theta_r \sim 2(1-r)^{\frac{1}{2}}$$

as $r \longrightarrow 1$. Thus,

$$\int_0^{2\pi} \left(1 - |S(re^{i\theta})|\right) d\theta \gtrsim (1-r)^{\frac{1}{2}}.$$

The conclusion $S' \notin B^{\frac{2}{3}}(\mathbb{D})$ comes from Corollary 6.5.

Theorem 6.9 (Ahern–Clark [4]) *Let ϕ be an inner function such that $\phi' \in B^{2/3}(\mathbb{D})$. Then ϕ is a Blaschke product.*

Proof. By Corollary 6.4 and Lemma 6.8, ϕ cannot have a singular inner factor.

As a matter of fact, there are Blaschke products satisfying $B' \in B^{2/3}(\mathbb{D})$. For example, finite Blaschke products have this property. Moreover, a family of infinite Blaschke products with $B' \in B^p(\mathbb{D})$, $0 < p < 1$, is given by Theorem 9.2.

6.4 A Construction of S with $S' \in B^p(\mathbb{D})$ for All $0 < p < \frac{2}{3}$

An immediate consequence of Theorem 6.9 is that if S is a nonconstant singular inner function, then

$$S' \notin B^{2/3}(\mathbb{D}).$$

However, the possibility of having a singular inner function S with

$$S' \in B^p(\mathbb{D}),$$

for some $0 < p < 2/3$, is not excluded. The following construction shows that this is indeed the case.

Lemma 6.10 *Let $0 < \alpha < 1$, $\beta > 2 - \alpha$, and $\sigma > 0$. Let*

$$\phi(z) = \frac{1}{(1-z)^\beta} \exp\left(\frac{-\sigma}{1-z}\right), \qquad (z \in \mathbb{D}).$$

Then

$$\int_0^{2\pi} |\phi(re^{i\theta})| \, d\theta \leq C_\alpha \frac{\sigma^{\alpha-1}}{(1-r)^{\frac{1+\alpha}{2}+\beta-2}}, \qquad (0 \leq r < 1).$$

Proof. By definition,

$$|\phi(re^{i\theta})| = \frac{1}{|1-re^{i\theta}|^\beta} \exp\left(-\sigma \frac{1-r\cos\theta}{|1-re^{i\theta}|^2}\right).$$

Since $e^{-x} < x^{-(1-\alpha)}$, we obtain

$$|\phi(re^{i\theta})| \leq \frac{1}{\sigma^{1-\alpha}} \frac{|1-re^{i\theta}|^{2(1-\alpha)-\beta}}{\left(1-r\cos\theta\right)^{1-\alpha}}.$$

If we integrate with respect to θ, and apply Hölder's inequality with $p = 1/(1-\alpha)$ and $q = 1/\alpha$, to the integral means

$$I_r = \int_0^{2\pi} |\phi(re^{i\theta})| \, d\theta,$$

we obtain the upper estimation

$$
\begin{aligned}
I_r &\leq \frac{1}{\sigma^{1-\alpha}} \int_0^{2\pi} \frac{1}{\left(1-r\cos\theta\right)^{1-\alpha}} \times \frac{1}{|1-re^{i\theta}|^{\beta-2(1-\alpha)}} \, d\theta \\
&\leq \frac{1}{\sigma^{1-\alpha}} \left(\int_0^{2\pi} \frac{d\theta}{1-r\cos\theta}\right)^{1-\alpha} \left(\int_0^{2\pi} \frac{d\theta}{|1-re^{i\theta}|^{1+\frac{\alpha+\beta-2}{\alpha}}}\right)^\alpha \\
&\leq \frac{1}{2\sigma^{1-\alpha}} \left(\frac{2\pi}{\sqrt{1-r^2}}\right)^{1-\alpha} \left(\frac{c_\alpha}{(1-r)^{\frac{\alpha+\beta-2}{\alpha}}}\right)^\alpha \\
&\leq C_\alpha \frac{\sigma^{\alpha-1}}{(1-r)^{\frac{1+\alpha}{2}+\beta-2}}.
\end{aligned}
$$

Theorem 6.11 (Allen–Belna [5]) *Let $\sigma_n \geq 0$, $n \geq 1$, and assume that*

$$\sum_{n=1}^\infty \sigma_n^\alpha < \infty$$

for some $0 < \alpha < 1$. Put

$$S(z) = \exp\left(-\sum_{n=1}^\infty \sigma_n \frac{e^{i\theta_n}+z}{e^{i\theta_n}-z}\right), \qquad (z \in \mathbb{D}),$$

where $(e^{i\theta_n})_{n\geq1}$ is an arbitrary sequence on $\mathbb{T}$. Then

$$S' \in \bigcap_{0<p<\frac{2}{3+\alpha}} B^p(\mathbb{D}).$$

Proof. Clearly, we have

$$S(0) = \exp\left(-\sum_{n=1}^{\infty} \sigma_n\right) > 0.$$

Hence,

$$S(z)S(0) = \exp\left(-\sum_{n=1}^{\infty} \sigma_n \frac{2e^{i\theta_n}}{e^{i\theta_n} - z}\right), \qquad (z \in \mathbb{D}),$$

which implies

$$|S(re^{i\theta})|\, S(0) = \exp\left(-\sum_{n=1}^{\infty} 2\sigma_n \frac{1 - r\cos(\theta - \theta_n)}{|e^{i\theta_n} - re^{i\theta}|^2}\right).$$

Since the terms in the summation are positive, the rough estimation

$$\exp\left(-\sum_{k=1}^{\infty} 2\sigma_k \frac{1 - r\cos(\theta - \theta_k)}{|e^{i\theta_k} - re^{i\theta}|^2}\right) \leq \exp\left(-2\sigma_n \frac{1 - r\cos(\theta - \theta_n)}{|e^{i\theta_n} - re^{i\theta}|^2}\right)$$

is valid for any arbitrary choice of $n \geq 1$. Hence, for any $n \geq 1$,

$$|S(z)| \leq \frac{1}{S(0)} \left|\exp\left(-2\sigma_n \frac{e^{i\theta_n}}{e^{i\theta_n} - z}\right)\right|. \qquad (6.1)$$

As a special case of (4.9),

$$\frac{S'(z)}{S(z)} = -\sum_{n=1}^{\infty} \sigma_n \frac{2e^{i\theta_n}}{(e^{i\theta_n} - z)^2}, \qquad (z \in \mathbb{D}).$$

Hence, by (6.1),

$$|S'(z)| \leq 2\sum_{n=1}^{\infty} \sigma_n \frac{|S(z)|}{|e^{i\theta_n} - z|^2}$$

$$\leq \frac{2}{S(0)} \sum_{n=1}^{\infty} \sigma_n \left|\frac{1}{(e^{i\theta_n} - z)^2} \exp\left(-2\sigma_n \frac{e^{i\theta_n}}{e^{i\theta_n} - z}\right)\right|.$$

Therefore, by Lemma 6.10,

$$\int_0^{2\pi} |S'(re^{i\theta})| \, d\theta \lesssim \frac{\sum_{n=1}^\infty \sigma_n^\alpha}{(1-r)^{\frac{1+\alpha}{2}}},$$

and

$$\|S'\|_{B^p} = \int_0^1 \int_0^{2\pi} |S'(re^{i\theta})| \, (1-r)^{\frac{1}{p}-2} \, r \, dr \, d\theta$$

$$\lesssim \sum_{n=1}^\infty \sigma_n^\alpha \int_0^1 \frac{dr}{(1-r)^{\frac{5+\alpha}{2}-\frac{1}{p}}} \lesssim \sum_{n=1}^\infty \sigma_n^\alpha,$$

provided that $\frac{5+\alpha}{2} - \frac{1}{p} < 1$. This is equivalent to $p < \frac{2}{3+\alpha}$.

The conditions of Lemma 6.8 and Theorem 6.11 are clearly fulfilled by a singular measure consisting of a finite number of Dirac measures. Hence, we easily obtain the following result.

Corollary 6.12 *Let* $\sigma_1, \sigma_2, \ldots, \sigma_n > 0$, *and let* $\zeta_1, \zeta_2, \ldots, \zeta_n \in \mathbb{T}$. *Put*

$$S(z) = \exp\left(-\sum_{k=1}^n \sigma_k \frac{\zeta_k + z}{\zeta_k - z} \right), \qquad (z \in \mathbb{D}).$$

Then

$$S' \in \bigcap_{0 < p < 2/3} B^p(\mathbb{D}),$$

but

$$S' \notin B^{2/3}(\mathbb{D}).$$

6.5 A Generalized Cantor Set

Let E be a compact subset of $[0,1]$. Given $\varepsilon > 0$, define

$$E_\varepsilon = \{ x \in [0,1] : \ \mathrm{dist}(x, E) \leq \varepsilon \}.$$

We say that E is of *type* β, $0 < \beta \leq 1$, if

$$|E_\varepsilon| \lesssim \varepsilon^\beta$$

as $\varepsilon \longrightarrow 0$. We remind that $|E_\varepsilon|$ refers to the Lebesgue measure of E_ε.

Lemma 6.13 *There is a continuous singular measure* σ *on* $[0,1]$ *whose support is a set of type* β *for every* $\beta < 1$.

Proof. The Cantor function built on a Cantor set gives a continuous singular measure. Hence, it is enough to pick a proper generalized Cantor set. See Sect. 2.3.

Fix $\rho \in (0,1)$. Put $I_0 = [0,1]$. Then $(I_n)_{n \geq 1}$ is defined similar to the classical construction. The set I_n consists of 2^n intervals, each of length ρ^{n^2}, and I_{n+1} is obtained from I_n by deleting open intervals from the middle of each interval of I_n, so that I_{n+1} consists of 2^{n+1} intervals, each of length $\rho^{(n+1)^2}$. Then we define

$$E = \bigcap_{n=0}^{\infty} I_n.$$

Now, given $\varepsilon > 0$, choose n such that

$$\rho^{(n+1)^2} \leq \varepsilon < \rho^{n^2}. \tag{6.2}$$

Since $E \subset I_n$, we have $E_\varepsilon \subset (I_n)_\varepsilon$. That the set I_n consists of 2^n interval of length $\rho^{n^2} > \varepsilon$ ensures that $(I_n)_\varepsilon$ is contained in the union of 2^n intervals of length $3\rho^{n^2}$. Hence,

$$|E_\varepsilon| \leq 3 \times 2^n \rho^{n^2}.$$

But, according to (6.2),

$$n < \sqrt{\frac{\log \varepsilon}{\log \rho}} \leq n + 1.$$

Therefore,

$$|E_\varepsilon| \leq c\varepsilon \exp\left(c' \sqrt{-\log \varepsilon} \right),$$

where c and c' are some positive constants (depending on ρ, but not on ε). Finally, note that, given $\beta < 1$, the relation

$$\varepsilon \exp\left(c' \sqrt{-\log \varepsilon} \right) \leq \varepsilon^\beta$$

holds for all small values of ε.

A similar construction works for the subsets of $\mathbb{T}$.

6.6 An Example of S with $S' \in B^p(\mathbb{D})$ for All $0 < p < \frac{2}{3}$

In Theorem 6.11, we dealt with singular inner functions constructed with purely atomic measures. Then, as a special case, Corollary 6.12 says that $S' \in B^p(\mathbb{D})$, for all $0 < p < 2/3$, where S is any singular inner function whose measure consists of a finite number of atoms. In this section, we discuss a similar phenomenon. But, our measures are not necessarily atomic measures and we could also apply the result to certain continuous singular measures. The following result can also be viewed as a strengthened version of Theorem 6.1.

Theorem 6.14 (Ahern–Clark [5]) *Let σ be a positive singular measure on $\mathbb{T}$ which is carried on a compact set of type β, for some $\beta \in (0,1]$. Let S be the corresponding singular inner function. Then*

$$\int_{\mathbb{T}} (1 - |S(re^{i\theta})|)\, d\theta \lesssim (1-r)^q$$

for all $q < \frac{\beta}{2}$. In particular,

$$S' \in \bigcap_{0 < p < \frac{2}{4-\beta}} B^p(\mathbb{D}).$$

Proof. Put

$$P[\sigma](re^{i\theta}) = \int_{\mathbb{T}} \frac{1 - r^2}{|e^{it} - re^{i\theta}|^2}\, d\sigma(e^{i\theta}), \qquad (re^{i\theta} \in \mathbb{D}).$$

Then

$$P[\sigma](re^{i\theta}) \leq 4 \int_{\mathbb{T}} \frac{1 - r^2}{|e^{it} - e^{i\theta}|^2}\, d\sigma(e^{i\theta}) \leq 8\|\sigma\| \frac{1-r}{d^2(e^{i\theta})},$$

where

$$d(e^{i\theta}) = \text{dist}(e^{i\theta}, E) = \inf_{e^{it} \in E} |e^{it} - e^{i\theta}|.$$

Since

$$|S(re^{i\theta})| = \exp\left(-P[\sigma](re^{i\theta})\right)$$

and $1 - e^{-x} \leq x$, $x \geq 0$, we conclude that the estimation

$$1 - |S(re^{i\theta})| \leq \min\left\{1, 8\|\sigma\| \frac{1-r}{d^2(e^{i\theta})}\right\}, \qquad (re^{i\theta} \in \mathbb{D}), \qquad (6.3)$$

holds.

Note that our assumption on the type of E means that

$$\left|\left\{ e^{i\theta} \in \mathbb{T} : d(e^{i\theta}) \leq \varepsilon \right\}\right| \lesssim \varepsilon^{\beta} \qquad (6.4)$$

as $\varepsilon \longrightarrow 0$.

Fix $q < \frac{\beta}{2}$. Let $(\gamma_n)_{n \geq 0}$ be any decreasing sequence of real numbers and put

$$I_0 = \left\{ e^{i\theta} \in \mathbb{T} : d(e^{i\theta}) \leq 2(1-r)^{\gamma_0} \right\}$$

and, for $n \geq 1$,

$$I_n = \left\{ e^{i\theta} \in \mathbb{T} : 2(1-r)^{\gamma_{n-1}} < d(e^{i\theta}) \leq 2(1-r)^{\gamma_n} \right\}.$$

Hence, by (6.4) and (6.3),

$$\int_{I_0} (1 - |S(re^{i\theta})|) \, d\theta \leq |I_0| \lesssim (1 - r)^{\beta\gamma_0}$$

and similarly, for $n \geq 1$,

$$\int_{I_n} (1 - |S(re^{i\theta})|) \, d\theta \lesssim \frac{1 - r}{(1 - r)^{2\gamma_{n-1}}} \, |I_n| \lesssim (1 - r)^{1 - 2\gamma_{n-1} + \beta\gamma_n}.$$

Hence, if $(\gamma_n)_{n \geq 0}$ is such that

$$\beta\gamma_0 = 1 - 2\gamma_{n-1} + \beta\gamma_n = q, \qquad (n \geq 1),$$

then we have

$$\int_{\cup_{n=0}^N I_n} (1 - |S(re^{i\theta})|) \, d\theta \lesssim N \, (1 - r)^q. \qquad (6.5)$$

Two comments are in order. First, the sequence $(\gamma_n)_{n \geq 0}$ defined by

$$\gamma_0 = \frac{q}{\beta} \qquad \text{and} \qquad \gamma_n = \frac{2\gamma_{n-1} + q - 1}{\beta}, \qquad (n \geq 1),$$

is decreasing. This can be easily verified via relations

$$\gamma_0 - \gamma_1 = \frac{\beta - 2q}{\beta^2} \qquad \text{and} \qquad \gamma_n - \gamma_{n-1} = \frac{2(\gamma_{n-1} - \gamma_{n-2})}{\beta}, \qquad (n \geq 1).$$

Second, the unique solution is also given by

$$\left(\frac{\beta}{2}\right)^n \gamma_n = \frac{q}{\beta} - \frac{1 - q}{2} \left(1 + \left(\frac{\beta}{2}\right) + \left(\frac{\beta}{2}\right)^2 + \cdots + \left(\frac{\beta}{2}\right)^{n-1}\right), \qquad (n \geq 1).$$

As $n \longrightarrow \infty$, the right side converges to

$$\frac{q}{\beta} - \frac{1 - q}{2} \frac{1}{1 - \frac{\beta}{2}} = \frac{2q - \beta}{\beta(2 - \beta)} < 0.$$

Hence, there is an N_0 such that $\gamma_{N_0} < 0$. Therefore, $\mathbb{T} = \cup_{n=0}^{N_0} I_n$, and by (6.5), we get

$$\int_{\mathbb{T}} (1 - |S(re^{i\theta})|) \, d\theta \lesssim (1 - r)^q.$$

By Corollary 6.6, this estimation implies $S' \in B^p(\mathbb{D})$, for any $0 < p < \frac{1}{2-q}$. Thus, $S' \in B^p(\mathbb{D})$, for all $0 < p < \frac{2}{4-\beta}$.

Theorem 6.14 and Lemma 6.8 together imply the following result.

Corollary 6.15 *Let σ be a positive singular measure on $\mathbb{T}$ whose support is of type β, for all $\beta < 1$. Let S be the corresponding singular inner function.*

Then

$$S' \in \bigcap_{0<p<\frac{2}{3}} B^p(\mathbb{D}),$$

but

$$S' \notin B^{2/3}(\mathbb{D}).$$

In particular, if σ consists of a finite number of Dirac measures, then the preceding result applies and hence Corollary 6.12 can be considered as a special case of this result. Moreover, in the light of Lemma 6.13, we also have the following version.

Corollary 6.16 *There is a singular inner function S, which is constructed by a continuous singular measure, and such that*

$$S' \in \bigcap_{0<p<\frac{2}{3}} B^p(\mathbb{D}),$$

but

$$S' \notin B^{2/3}(\mathbb{D}).$$

Chapter 7
The Derivative of a Blaschke Product

7.1 Frostman's Theorem, Local Version

Let $(z_n)_{n\geq 1}$ be a Blaschke sequence and let

$$B(z) = \prod_{n=1}^{\infty} \frac{|z_n|}{z_n} \frac{z_n - z}{1 - \bar{z}_n z}.$$

For a fixed point $z \in \mathbb{D}$, we know that the partial products

$$B_N(z) = \prod_{n=1}^{N} \frac{|z_n|}{z_n} \frac{z_n - z}{1 - \bar{z}_n z}$$

converge to $B(z)$. Indeed, more is true. The convergence is uniform on each compact subset of an open set Ω which contains the open unit disc. Naturally, we may ask a similar question for the behavior of $B_N(e^{i\theta})$ as $N \longrightarrow \infty$. Does the limit

$$\lim_{N\to\infty} B_N(e^{i\theta}) \tag{7.1}$$

necessarily exist? This question is more interesting to consider when $e^{i\theta}$ is a point of accumulation of the zeros of B. Since otherwise, we know that B is in fact analytic at this point and the convergence of B_N is even uniform on a small disc around $e^{i\theta}$. In the general case, the answer is not affirmative. For example, if all the zeros are on the interval $[0, 1)$, then

$$B_N(z) = \prod_{n=1}^{N} \frac{r_n - z}{1 - r_n z},$$

which shows that $B_N(1) = (-1)^N$, and thus $\lim_{N\to\infty} B_N(1)$ does not exist.

Let us look at the boundary values of a Blaschke product from a different point of view. According to Fatou's theorem, we know that the radial limits

J. Mashreghi, *Derivatives of Inner Functions*, Fields Institute Monographs 31, 99
DOI 10.1007/978-1-4614-5611-7_7, © Springer Science+Business Media New York 2013

$\lim_{r \to 1} B(re^{i\theta})$ exist for almost all $e^{i\theta} \in \mathbb{T}$. But, first, the exceptional set of measure zero on which the radial limits fail to exist depends on B. Second, even for a fixed B, we do not have any intrinsic description of this set, except of course the fact that its Lebesgue measure is zero. Hence, given B and a fixed point $e^{i\theta} \in \mathbb{T}$, Fatou's result does not tell us anything about the existence of the radial limit at this point. The following theorem is a significant result which partially answers this question. This theorem gives a sufficient condition under which a Blaschke product has a unimodular radial limit at a given point on $\mathbb{T}$. We naturally expect this value to be the limit of partial product at the boundary point.

For each Blaschke sequence $(z_n)_{n \geq 1}$, any subsequence $(z_{n_k})_{k \geq 1}$ (finite or infinite) of $(z_n)_{n \geq 1}$ is by itself a Blaschke sequence and thus the Blaschke product

$$B(z; \{z_{n_k}\}) = \prod_k \frac{|z_{n_k}|}{z_{n_k}} \frac{z_{n_k} - z}{1 - \bar{z}_{n_k} z}$$

is well-defined. The Blaschke products $B(z; \{z_{n_k}\})$ are called the *subproducts* of B. Note that B is a subproduct of itself corresponding to the whole sequence $(z_n)_{n \geq 1}$. For a more complete version of the following result, see Theorem 7.3.

Theorem 7.1 (Frostman [23]) *Let $(z_n)_{n \geq 1}$ be a Blaschke sequence, and fix $e^{i\theta} \in \mathbb{T}$. Suppose that*

$$\sum_{n=1}^{\infty} \frac{1 - |z_n|}{|e^{i\theta} - z_n|} < \infty. \tag{7.2}$$

Then each subproduct $B(z; \{z_{n_k}\})$ is convergent at the boundary point $z = e^{i\theta}$ and, moreover,

$$\lim_{r \to 1} B(re^{i\theta}; \{z_{n_k}\}) = B(e^{i\theta}; \{z_{n_k}\}) = \lim_{N \to \infty} B_N(e^{i\theta}; \{z_{n_k}\}).$$

Conversely, if all subproducts of B have unimodular radial limits at $e^{i\theta}$ then (7.2) holds.

Proof. It is enough to show that the partial products

$$B_N(z) = \prod_{n=1}^{N} \frac{|z_n|}{z_n} \frac{z_n - z}{1 - \bar{z}_n z}, \qquad (z \in \mathbb{D}),$$

are uniformly convergent on the closed ray $[0, e^{i\theta}]$. Write $B = \prod b_{z_n}$. Then, according to (1.14), we have

$$b_{z_n}(z) = 1 - (1 - |z_n|) \left(1 + \frac{|z_n|(1 + |z_n|)z}{z_n(1 - \bar{z}_n z)}\right)$$

and thus

$$|1 - b_{z_n}(re^{i\theta})| \le (1 - |z_n|) + 2\,\frac{1 - |z_n|}{|1 - re^{-i\theta}z_n|}.$$

Thus, for all $r \in [0, 1]$,

$$|1 - b_{z_n}(re^{i\theta})| \le (1 - |z_n|) + 4\,\frac{1 - |z_n|}{|e^{i\theta} - z_n|}.$$

The assumption (7.2) now ensures that the partial products B_N are uniformly convergent on the ray $[0, e^{i\theta}]$. Since each B_N is a continuous function of r, for $r \in [0, 1]$, the limiting function B is also continuous on the ray $[0, e^{i\theta}]$. This means that B is convergent at the point $e^{i\theta}$, B has a radial limit at $e^{i\theta}$, and, moreover, its radial limit at this point is exactly $B(e^{i\theta})$. If (7.2) holds, clearly it also holds for any subsequence of $(z_n)_{n\ge 1}$. Thus, any subproduct of B also has a unimodular radial limit.

Now, suppose that B and all its subproducts have a unimodular radial limit at $e^{i\theta}$. Since B is a bounded function, $B(z)$ has the same unimodular limit as z tends to $e^{i\theta}$ from within a Stolz domain anchored at $e^{i\theta}$. Fix one of these domains, say with opening $\pi/2$, and without loss of generality assume that $e^{i\theta} = 1$. Therefore, since B has a non-tangential unimodular limit, there are a finite number of zeros $(z_n)_{n\ge 1}$ in this Stolz domain.

We consider three subproduct of B. The first subproduct B_1 is formed with zeros in the Stolz domain and zeros in the second or third quadrant. Since the first subproduct has a finite number of zeros in the Stolz domain, this subproduct is indeed analytic at the point 1, and moreover

$$\sum_{z_n \in B_1^{-1}(0)} \frac{1 - |z_n|}{|1 - z_n|} \le C \sum_{z_n \in B_1^{-1}(0)} (1 - |z_n|) < \infty.$$

The second subproduct B_2 is formed with zeros in the first quadrant which are outside the Stolz domain, i.e. the zeros which satisfy

$$\frac{\Im z_n}{|1 - z_n|} \ge \frac{1}{\sqrt{2}}. \tag{7.3}$$

For $0 \le r < 1$, we have

$$\arg B_2(r) = \sum_{z_n \in B_2^{-1}(0)} \arcsin \frac{r\,\Im(z_n)\,(1 - |z_n|^2)}{|z_n|\,|z_n - r|\,|1 - \bar{z}_n r|}.$$

Note that the factor $1 - |z_n|$ assures the convergence of the series for each fixed r, and moreover, it implies that $\arg B_2$ is a continuous function of r on the interval $[0, 1)$. Since we assume that B_2 has a unimodular limit at the point 1, the limit

$$\lim_{r \to 1^-} \arg B_2(r)$$

has to exist, and thus we can consider $\arg B_2$ as a continuous function on $[0, 1]$. Hence, $\arg B_2(r)$ is uniformly bounded on $[0, 1]$, say

$$\sum_{z_n \in B_2^{-1}(0)} \arcsin \frac{r \, \Im(z_n) \, (1 - |z_n|^2)}{|z_n| \, |z_n - r| \, |1 - \bar{z}_n r|} \leq C,$$

and, in particular, we have

$$\sum_{z_n \in B_2^{-1}(0)} \arcsin \frac{\Im(z_n) \, (1 - |z_n|^2)}{|z_n| \, |z_n - 1| \, |1 - \bar{z}_n|} \leq C.$$

Using (7.3), and that $\arcsin \theta \geq \theta$, we get

$$\sum_{z_n \in B_2^{-1}(0)} \frac{1 - |z_n|}{|1 - z_n|} \leq C.$$

The treatment of the third subproduct formed with zeros in the forth quadrant and outside the Stolz domain is similar. We thus obtain (7.2).

7.2 The Radial Variation

Let f be an analytic function on $\mathbb{D}$. The radial variation of f at $e^{i\theta}$ is defined to be

$$V(f, e^{i\theta}) = \int_0^1 |f'(re^{i\theta})| \, dr.$$

Geometrically speaking, $V(f, e^{i\theta})$ is the length of curve

$$[0, 1) \longrightarrow \mathbb{C}$$
$$r \longmapsto f(re^{i\theta}).$$

The following elementary result establishes the relation between the radial variation and the radial limit of f at $e^{i\theta}$.

Lemma 7.2 *Let $f : \mathbb{D} \longrightarrow \mathbb{C}$ be analytic, and let $e^{i\theta} \in \mathbb{T}$ be such that*

$$V(f, e^{i\theta}) < \infty.$$

Then

$$\lim_{r \to 1} f(re^{i\theta})$$

exists.

Proof. Since

$$f(\rho_2 e^{i\theta}) - f(\rho_1 e^{i\theta}) = \int_{\rho_1 e^{i\theta}}^{\rho_2 e^{i\theta}} f'(z)\, dz,$$

and thus

$$|f(\rho_2 e^{i\theta}) - f(\rho_1 e^{i\theta})| \leq \int_{\rho_1}^{\rho_2} |f'(re^{i\theta})|\, dr,$$

the assumption $V(f, e^{i\theta}) < \infty$ ensures that the family $\big(f(re^{i\theta})\big)_{0 \leq r < 1}$ is a Cauchy net, as $r \longrightarrow 1$, and thus the limit $\lim_{r \to 1} f(re^{i\theta})$ exists.

The inverse of Lemma 7.2 is far from being true. W. Rudin extensively studied this phenomenon [42]. Among other things, he showed that there is a Blaschke product B for which $V(B, e^{i\theta}) = \infty$ for almost all $e^{i\theta} \in \mathbb{T}$. However, J. Bourgain showed that, for each $f \in H^\infty(\mathbb{D})$, the set

$$\{e^{i\theta} \in \mathbb{T} : V(f, e^{i\theta}) < \infty\}$$

is nonempty and in fact of Hausdorff dimension 1 [10]. Note that at the same time, by Fatou's theorem, B has radial limit almost everywhere on $\mathbb{T}$.

Let $(z_n)_{n \geq 1}$ be a Blaschke sequence, and fix $e^{i\theta} \in \mathbb{T}$. According to Theorem 7.1, if

$$\sum_{n=1}^{\infty} \frac{1 - |z_n|}{|e^{i\theta} - z_n|} = \infty, \tag{7.4}$$

then one of the subproducts of B does not have a unimodular radial limit. This means that either the radial limit does not exist, or it exists but it is not unimodular. For example, if we consider the Blaschke product formed with $z_n = 1 - n^{-2}$, $n \geq 1$, then, on the one hand (7.4) is fulfilled, and on the other hand, direct verifications show that $\lim_{r \to 1} B(r) = 0$. Hence, B itself can be considered as the subproduct whose existence is manifested in Theorem 7.1. However, more can be said in this regard. As a matter of fact, under the condition (7.4), there is always a subproduct which does not have the radial limit at $e^{i\theta}$.

Theorem 7.3 (Frostman [23], Cargo [12]) *Let $(z_n)_{n \geq 1}$ be a Blaschke sequence, and let $e^{i\theta} \in \mathbb{T}$. Then the following are equivalent:*

(i)

$$\sum_{n=1}^{\infty} \frac{1 - |z_n|}{|e^{i\theta} - z_n|} < \infty;$$

(ii) The radial variations of all subproducts of B at $e^{i\theta}$ are uniformly bounded, i.e. there is a constant C such that, for each subsequence $(z_{n_k})_{k \geq 1}$,

$$V\big(B(z; \{z_{n_k}\}), e^{i\theta}\big) \leq C;$$

(iii) Each subproduct of B has a radial limit at $e^{i\theta}$;
(iv) Each subproduct of B has a unimodular radial limit at $e^{i\theta}$.

Proof. Theorem 7.1 says $(i) \Longleftrightarrow (iv)$. Hence, we show that $(i) \Longrightarrow (ii) \Longrightarrow (iii) \Longrightarrow (i)$.

$(i) \Longrightarrow (ii)$: By (4.2), we have

$$|B'(z; \{z_{n_k}\})| \leq \sum_{k=1}^{\infty} \frac{1 - |z_{n_k}|^2}{|1 - \bar{z}_{n_k} z|^2}.$$

Hence, if we integrate on the radius $[0, e^{i\theta})$, we obtain

$$V\big(B(z; \{z_{n_k}\}), e^{i\theta}\big) \leq \frac{\pi}{2} \sum_{k=1}^{\infty} \frac{1 - |z_{n_k}|^2}{|e^{i\theta} - z_{n_k}|} \leq \pi \sum_{n=1}^{\infty} \frac{1 - |z_n|}{|e^{i\theta} - z_n|} < \infty.$$

$(ii) \Longrightarrow (iii)$: This follows from Lemma 7.2.

$(iii) \Longrightarrow (i)$: This is equivalent to assume that (7.4) holds and then show some subproduct of B fails to have a radial limit at $e^{i\theta}$. Since

$$\frac{1 - |z_n|}{|e^{i\theta} - z_n|} = \frac{1 - |e^{-i\theta} z_n|}{|1 - e^{-i\theta} z_n|},$$

and

$$B(z; \{z_{n_k}\}) = \prod_k \frac{|z_{n_k}|}{z_{n_k}} \frac{z_{n_k} - z}{1 - \bar{z}_{n_k} z}$$

$$= \prod_k \frac{|e^{-i\theta} z_{n_k}|}{e^{-i\theta} z_{n_k}} \frac{e^{-i\theta} z_{n_k} - e^{-i\theta} z}{1 - \overline{e^{-i\theta} z_{n_k}} \, e^{-i\theta} z}$$

$$= B(e^{-i\theta} z; \{e^{-i\theta} z_{n_k}\}),$$

without loss of generality we may suppose that $\theta = 0$.

Since we assume that

$$\sum_{n=1}^{\infty} \frac{1 - |z_n|}{|1 - z_n|} = \infty,$$

there are three possibilities:

(i)

$$\sum_{\Im z_n > 0} \frac{1 - |z_n|}{|1 - z_n|} = \infty;$$

(ii)

$$\sum_{\Im z_n < 0} \frac{1 - |z_n|}{|1 - z_n|} = \infty;$$

(iii) There are infinitely many zeros on the radius $[0, 1)$.

We study the case (i) in detail. Since $B(r; \{z_n\}) = \overline{B(r; \{\bar{z}_n\})}$, the second case follows from the first one. Finally, a small modification of the end of proof of the case (i) yields a proof for the last case.

Let $E = \{z_n : 0 < \arg z_n < \frac{\pi}{2}\}$. The heart of the proof is the following fact: the assumption

$$\sum_{z_n \in E} \frac{1 - |z_n|}{|1 - z_n|} = \infty \tag{7.5}$$

implies that $B(z; E)$ carries the radius $[0, 1)$ onto a curve that rotates counterclockwise infinitely many times around the origin. Hence, we would be able to extract a subproduct whose radial limit at 1 either does not exist or exists, but it is zero. In the latter case (which can also be applied to the case (iii)) we will kick out more zeros in order to obtain a subproduct that fails to have a radial limit.

Write $z_n = r_n e^{i\theta_n}$. Since $|1 - z_n| \asymp (1 - r_n) + |\theta_n|$, our assumption (7.5) is equivalent to

$$\sum_{z_n \in E} \frac{1 - r_n}{\theta_n} = \infty.$$

For each $z_n \in E$, we have

$$\arg b_{z_n}(0) = 2 \arctan\left(\frac{(1 - r_n)}{(1 + r_n)\tan(\frac{\theta_n}{2})}\right) \geq 2 \arctan\left(\frac{1 - r_n}{\theta_n}\right).$$

Since $\arctan x \sim x$, as $x \longrightarrow 0$, we deduce that

$$\sum_{z_n \in E} \arg b_{z_n}(0) = \infty.$$

But,

$$\arg B(r; E) = \sum_{z_n \in E} \arg b_{z_n}(r), \qquad (0 \leq r \leq 1),$$

and thus

$$\lim_{r \to 1} \arg B(r; E) = \infty.$$

Therefore, either $B(z; E)$ does not have a radial limit at 1, or it has the radial limit zero. In the former case, the proof is finished. In the latter, we still need to dig and extract a subproduct $B(z; F)$ out of the subproduct $B(z; E)$, which fails to have a radial limit. We do it such a way that

$$\sum_{z_n \in F} \frac{1 - r_n}{\theta_n} = \infty$$

still holds, and thus $B(r; F)$ also winds counterclockwise around the origin, but $|B(r; F)| \longrightarrow 1$, at least for a sequence $r = \rho_k$ with $\rho_k \longrightarrow 1$.

For simplicity of notations, reindex the elements of E from 1 to ∞. For each integer n, let $s(n)$ to be the first integer bigger that k such that

$$\sum_{k=n}^{s(n)} \frac{1 - r_k}{\theta_k} \geq 1. \tag{7.6}$$

Put

$$F_n = \{ z_k : n \leq k \leq s(n) \}, \qquad (n \geq 1).$$

Let $n_1 = 1$. Since $|B(z, F_{n_1})| \longrightarrow 1$, as $|z| \longrightarrow 1$, we can choose ρ_1 such that

$$|B(r, F_{n_1})| \geq 1 - \frac{1}{2}, \qquad (\rho_1 \leq r \leq 1).$$

For any $n_2 > s(n_1)$, we have

$$|B(z, F_{n_2})| \geq 1 - 2 \frac{1 + |z|}{1 - |z|} \sum_{k=n_2}^{s(n_2)} (1 - |z_k|).$$

Hence, we can choose n_2 large enough such that

$$|B(r, F_{n_2})| \geq 1 - \frac{1}{2^2}, \qquad (0 \leq r \leq \rho_1).$$

Having found n_2, we now choose $\rho_2 > \rho_1$ such that

$$|B(r, F_{n_1} \cup F_{n_2})| \geq 1 - \frac{1}{2^2}, \qquad (\rho_2 \leq r \leq 1).$$

Continuing this process, we obtain two increasing sequences $(n_k)_{k \geq 1}$ and $(\rho_k)_{k \geq 1}$ such that, for each $k \geq 1$,

$$|B(r, F_{n_k})| \geq 1 - \frac{1}{2^k}, \qquad (0 \leq r \leq \rho_{k-1}), \tag{7.7}$$

and

$$|B(r, F_{n_1} \cup F_{n_2} \cup \cdots \cup F_{n_k})| \geq 1 - \frac{1}{2^k}, \qquad (\rho_k \leq r \leq 1). \tag{7.8}$$

Put

$$F = \bigcup_{k=1}^{\infty} F_{n_k}.$$

According to (7.6), we have

$$\sum_{z_n \in F} \frac{1 - r_n}{\theta_n} = \infty.$$

Moreover, by (7.7) and (7.8),

$$|B(\rho_k, F)| \geq \prod_{j=k}^{\infty} \left(1 - \frac{1}{2^k}\right), \qquad (k \geq 1).$$

Therefore, $|B(\rho_k; F)| \longrightarrow 1$, as $k \longrightarrow \infty$.

In case (iii), a small modification of the process above gives a subproduct $B(z, F)$ such that $|B(r; F)| \longrightarrow 1$, as $r \longrightarrow 1$ on a particular sequence ρ_k, $k \geq 1$, and at the same time $|B(r; F)| \longrightarrow 0$, as $r \longrightarrow 1$ through the points of F. Hence, the radial limit does not exist.

Certain immediate consequences of Theorem 7.3 are highlighted below.

(i) If all subproducts of a Blaschke product have radial limits at a boundary point, then these radial limits are all of modulus one.

(ii) If all subproducts of a Blaschke product have a finite radial variation at a boundary point, then these radial variations are uniformly bounded away from infinity.

(iii) Some subproduct of a Blaschke product has infinite radial variation at a point if and only if some subproduct (not necessarily the same) fails to have a radial limit at that boundary point.

7.3 Frostman's Theorems, Global Version

Fatou's theorem says that a bounded analytic function has radial limits almost everywhere on $\mathbb{T}$. Riesz's theorem completes the picture and says that the radial limits are unimodular almost everywhere on $\mathbb{T}$. If we put a stronger condition on the rate of growth of zeros of B, we naturally expect to get better results. In this section, we study some results of this type.

Theorem 7.4 (Frostman [23]) *Let $(z_n)_{n \geq 1}$ be a Blaschke sequence satisfying the stronger condition*

$$\sum_{n=1}^{\infty} (1 - |z_n|)^{\alpha} < \infty$$

for some $0 < \alpha < 1$, and let B be the Blaschke product formed with this sequence. Then B converges at all points of $\mathbb{T}$ and has a unimodular radial limit at these points except possibly on a set of α-capacity zero.

Proof. Let $E \subset \mathbb{T}$ be the set of all points $e^{i\theta}$ such that

$$\sum_{n=1}^{\infty} \frac{1 - |z_n|}{|e^{i\theta} - z_n|} = \infty. \tag{7.9}$$

According to Theorem 7.1, at all points of $\mathbb{T} \setminus E$, B converges and has a unimodular radial limit. Hence, it is enough to show that $C_\alpha(E) = 0$. (However, it is worthwhile to mention that B still may converge and have a unimodular limit at some points of E.)

Suppose, on the contrary, that $C_\alpha(E) > 0$ and we seek a contradiction. Thus, by Theorem 2.3, there is a positive Borel measure μ, $\mu \not\equiv 0$, whose support is in E, and a positive constant C, such that

$$P_{\mu,\alpha}(z) = \int_{\mathbb{C}} \frac{d\mu(e^{i\theta})}{|e^{i\theta} - z|^\alpha} \leq C, \qquad (z \in \mathbb{C}).$$

Let

$$I = \int_{\mathbb{C}} \left(\sum_{n=1}^{\infty} \frac{1 - |z_n|}{|e^{i\theta} - z_n|} \right) d\mu(e^{i\theta}).$$

Since $1 - |z_n| \leq |e^{i\theta} - z_n|$, then, on the one hand, we have

$$I = \sum_{n=1}^{\infty} (1 - |z_n|)^\alpha \int_{\mathbb{C}} \left(\frac{1 - |z_n|}{|e^{i\theta} - z_n|} \right)^{1-\alpha} \frac{d\mu(e^{i\theta})}{|e^{i\theta} - z_n|^\alpha}$$

$$\leq \sum_{n=1}^{\infty} (1 - |z_n|)^\alpha \int_{\mathbb{C}} \frac{d\mu(e^{i\theta})}{|e^{i\theta} - z_n|^\alpha}$$

$$= \sum_{n=1}^{\infty} (1 - |z_n|)^\alpha P_{\mu,\alpha}(z_n)$$

$$\leq C \sum_{n=1}^{\infty} (1 - |z_n|)^\alpha < \infty.$$

But, on the other hand, by (7.9),

$$I = \int_{\mathbb{C}} \left(\sum_{n=1}^{\infty} \frac{1 - |z_n|}{|e^{i\theta} - z_n|} \right) d\mu(e^{i\theta})$$

$$= \int_{E} \left(\sum_{n=1}^{\infty} \frac{1 - |z_n|}{|e^{i\theta} - z_n|} \right) d\mu(e^{i\theta}) = \infty.$$

This is a contradiction.

Corollary 7.5 (Cargo [12]) *Let* $(z_n)_{n \geq 1}$ *be a Blaschke sequence satisfying the stronger condition*

$$\sum_{n=1}^{\infty} (1 - |z_n|)^\alpha < \infty$$

for some $0 < \alpha < 1$, *and let* B *be the Blaschke product formed with this sequence. Then, for all* $e^{i\theta} \in \mathbb{T}$ *except possibly on a set of α-capacity zero, all*

the subproducts $B(z, \{z_{n_k}\})$ have finite radial variations, and for any such $e^{i\theta}$
the radial variations (corresponding to different subproducts) are uniformly
bounded away from infinity.

Remark. The upper bounds for the radial variations depend on θ, and thus
there might not exist a universal constant which works as an upper bound
for all such θ's.

Proof. The result follows from Theorem 7.4.

Example 7.6   Theorem 7.4 is almost sharp in the following sense. Given
a sequence of positive numbers $(r_n)_{n\geq 1}$ satisfying

$$\sum_{n=1}^{\infty}(1 - r_n) < \infty,$$

but, for some $0 < \alpha < 1$,

$$\sum_{n=1}^{\infty}(1 - r_n)^{\alpha} = \infty,$$

we are able to find a sequence of arguments $(\theta_n)_{n\geq 1}$ such that the Blaschke
product formed with zeros $(r_n e^{i\theta_n})_{n\geq 1}$ does not have a unimodular radial
limit on a set of positive $\beta-$capacity, for each $\beta < \alpha$.

Let E be the set constructed on $\mathbb{T}$ with

$$\ell_n = 2^{-n/\alpha}\, n^{-2/\alpha}$$

as described at the end of Sect. 2.3. The condition $2\ell_n < \ell_{n-1}$ is satisfied.
Hence, by (2.14), we have

$$\mu_{\beta}(E) = C \lim_{n\to\infty} 2^n\, \ell_n^{\beta} = C \lim_{n\to\infty} 2^{n(1-\frac{\beta}{\alpha})}\, n^{-2\beta/\alpha} = \infty$$

for each $\beta < \alpha$. Thus, by (2.11), we must also have

$$C_{\beta}(E) = \infty$$

for all $\beta < \alpha$. For simplicity, write $\varepsilon_n = 1 - r_n$. Then the relation

$$\sum_{n=1}^{\infty} \varepsilon_n^{\alpha} \leq \sum_{n=1}^{\infty} 2^n\, \varepsilon_{2^n}^{\alpha}$$

ensures that

$$\sum_{n=1}^{\infty} 2^n\, \varepsilon_{2^n}^{\alpha} = \infty,$$

and therefore, for infinitely many values of n, we must have

$$2^n \varepsilon_{2^n}^{\alpha} > \frac{1}{n^2}.$$

Equivalently, this means that, for infinitely many values of n,

$$\varepsilon_{2^n} > \ell_n. \tag{7.10}$$

Let θ_k be such that $z_k = r_k e^{i\theta_k}$ stays on the ray passing through the middle of I_k. Fix $e^{i\theta} \in E$. Hence, $e^{i\theta} \in I_k$ for some k with $2^{n-1} \le k < 2^n$. Then

$$|e^{i\theta} - z_k| \sin(\varphi_k) = |z_k| \sin(\phi_k) \le |z_k| \sin(\ell_n/2) \le \ell_n/2.$$

Therefore,

$$\sin(\varphi_k) \le \frac{\ell_n}{2|e^{i\theta} - z_k|} \le \frac{\ell_n}{2\varepsilon_k} \le \frac{\ell_n}{2\varepsilon_{2^n}}.$$

Hence, for infinitely many values of k, we have

$$\sin(\varphi_k) \le \frac{1}{2}.$$

This means that infinitely many zeros of B are in a Stolz domain with opening $\pi/6$ anchored at $e^{i\theta}$. Thus, B does not have a unimodular limit at this point.

Theorem 7.4 along with the preceding observation can be stated in the following technical language.

Corollary 7.7 *Let $(z_n)_{n \ge 1}$ be a Blaschke sequence, and let B be the Blaschke product formed with this sequence. Let $E \subset \mathbb{T}$ be the set of all points at which B does not converge or does not have a unimodular radial limit. If, for some $0 < \alpha < 1$,*

$$\sum_{n=1}^{\infty} (1 - |z_n|)^{\alpha} < \infty$$

then the Hausdorff dimension of E is at most α, and if

$$\sum_{n=1}^{\infty} (1 - |z_n|)^{\alpha} = \infty$$

then the Hausdorff dimension of E can be at least α.

We emphasize that in the second case of above corollary, the Hausdorff dimension *can* be at least α. But, it might be actually less than α. For example, if all the zeros are on the ray $[0, 1)$, then the exceptional set is precisely $E = \{1\}$ whose logarithmic capacity is zero, while given any $\alpha \in (0, 1)$ we can easily choose a sequence $(r_n)_{n \ge 1}$ such that

$$\sum_{n=1}^{\infty} (1 - r_n)^{\alpha} = \infty$$

and

$$\sum_{n=1}^{\infty}(1 - r_n)^{\alpha+\varepsilon} < \infty.$$

Hence, to avoid any ambiguity, it is better to say that there are Blaschke sequences $(z_n)_{n\geq 1}$ such that

$$\sum_{n=1}^{\infty}(1 - |z_n|)^{\alpha} = \infty$$

and the corresponding Blaschke product does not have unimodular limits on a set of Hausdorff dimension of at least α.

Since $C_1(\mathbb{T}) = 0$ Theorem 7.4 does not provide any information if $\alpha = 1$. However, a proper modification of its proof gives the following result.

Theorem 7.8 (Frostman [23]) *Let $(z_n)_{n\geq 1}$ be a Blaschke sequence satisfying the stronger condition*

$$\sum_{n=1}^{\infty}(1 - |z_n|) \log \frac{1}{(1 - |z_n|)} < \infty,$$

and let B be the Blaschke product formed with this sequence. Then B converges at all points of $\mathbb{T}$ and has a unimodular radial limit at these points except possibly on a set of Lebesgue measure zero.

Proof. It is enough to note that

$$\int_{\mathbb{T}}\left(\sum_{n=1}^{\infty}\frac{1 - |z_n|}{|e^{i\theta} - z_n|}\right) d\theta = \sum_{n=1}^{\infty}(1 - |z_n|)\int_{-\pi}^{\pi}\frac{d\theta}{|e^{i\theta} - z_n|}$$

$$= 2\sum_{n=1}^{\infty}(1 - |z_n|)\int_{0}^{\pi}\frac{dt}{\left|e^{it} - |z_n|\right|}$$

$$\leq C\sum_{n=1}^{\infty}(1 - |z_n|)\log\frac{1}{(1 - |z_n|)} < \infty.$$

Hence, the set of point $e^{i\theta} \in \mathbb{T}$ such that

$$\sum_{n=1}^{\infty}\frac{1 - |z_n|}{|e^{i\theta} - z_n|} = \infty$$

has the Lebesgue measure zero. Hence, by Theorem 7.1, B converges and has a unimodular radial limit almost everywhere on $\mathbb{T}$.

Corollary 7.9 *Let $(z_n)_{n \geq 1}$ be a Blaschke sequence satisfying the stronger condition*

$$\sum_{n=1}^{\infty} (1 - |z_n|) \log \frac{1}{(1 - |z_n|)} < \infty,$$

and let B be the Blaschke product formed with this sequence. Then, for all $e^{i\theta} \in \mathbb{T}$ except possibly on a set of Lebesgue measure zero, all the subproducts $B(z, \{z_{n_k}\})$ have finite radial variations, and for any such $e^{i\theta}$ these radial variations are uniformly bounded away from infinity.

Proof. The result follows from Theorems 7.3 and 7.8.

A simple calculation shows that

$$(1 - |z_n|) \log \frac{1}{(1 - |z_n|)} \leq 2(1 - |z_n|) \log n + \frac{2}{n^{3/2}}, \qquad (n \geq 1).$$

As a matter of fact, if $\log 1/(1 - |z_n|) \leq 2 \log n$, then the inequality trivially holds. And, if not, i.e. if $\log 1/(1 - |z_n|) > 2 \log n$, then

$$(1 - |z_n|) \log \frac{1}{(1 - |z_n|)} \leq 2(1 - |z_n|)^{3/4} \leq \frac{2}{n^{3/2}}.$$

Hence, in each case, the inequality is valid. Therefore, we obtain the following weaker version of Theorem 7.8.

Corollary 7.10 *Let $(z_n)_{n \geq 1}$ be a Blaschke sequence satisfying the stronger condition*

$$\sum_{n=1}^{\infty} (1 - |z_n|) \log n < \infty,$$

and let B be the Blaschke product formed with this sequence. Then B converges at all points of $\mathbb{T}$ and has a unimodular radial limit at these points except possibly on a set of Lebesgue measure zero.

Example 7.11   The Corollary 7.10 is sharp in the following sense. Given a sequence of positive numbers $(r_n)_{n \geq 1}$ satisfying

$$\sum_{n=1}^{\infty} (1 - r_n) < \infty,$$

but

$$\sum_{n=1}^{\infty} (1 - r_n) \log n = \infty,$$

we are able to find a sequence of arguments $(\theta_n)_{n \geq 1}$ such that, for each $e^{i\theta} \in \mathbb{T}$, the Blaschke product formed with zeros $z_n = r_n e^{i\theta_n}$, $n \geq 1$, or one of its subproducts do not have a unimodular radial limit at this point. In other words, according to Theorem 7.1, we should find the arguments θ_n, $n \geq 1$, such that

$$\sum_{n=1}^{\infty} \frac{1 - |z_n|}{|e^{i\theta} - z_n|} = \infty$$

for all $e^{i\theta} \in \mathbb{T}$. Hence, for $2^{n-1} \leq k < 2^n$, put

$$\theta_k = \frac{2\pi k}{2^{n-1}},$$

where $n \geq 1$. Since

$$|e^{i\theta} - z_k| \leq (1 - r_k) + |\theta - \theta_k|,$$

we see that

$$\begin{aligned}
S_n &= \sum_{k=2^{n-1}}^{2^n - 1} \frac{1 - |z_k|}{|e^{i\theta} - z_k|} \\
&\geq (1 - r_{2^n}) \sum_{k=1}^{2^{n-1}} \frac{1}{(1 - r_{2^{n-1}}) + \frac{2\pi k}{2^{n-1}}} \\
&= 2^{n-1}(1 - r_{2^n}) \sum_{k=1}^{2^{n-1}} \frac{1}{2^{n-1}(1 - r_{2^{n-1}}) + 2\pi k}.
\end{aligned}$$

As $n \longrightarrow \infty$, we know that $2^n(1 - r_{2^n}) \longrightarrow 0$. Therefore, for n large enough,

$$\begin{aligned}
S_n &\geq 2^{n-1}(1 - r_{2^n}) \times \frac{1}{4\pi} \sum_{k=1}^{2^{n-1}} \frac{1}{k} \\
&\geq \frac{2^{n-1}}{4\pi}(1 - r_{2^n}) \log(2^{n-1} - 1).
\end{aligned}$$

By assumption,

$$\sum_{n=1}^{\infty} n \, 2^n \, (1 - r_{2^n}) = \infty.$$

Therefore,

$$\sum_{n=1}^{\infty} \frac{1 - |z_n|}{|e^{i\theta} - z_n|} = \sum_{n=1}^{\infty} S_n = \infty.$$

7.4 An Example of B with $B' \notin \mathcal{N}$

The relations (4.16) and (4.18) provide a good basis to deceive us to conjecture that B' should be in a Hardy space $H^p(\mathbb{D})$, at least for a small p. However, on the contrary, O. Frostman [23] constructed a Blaschke product whose zeros satisfy even the stronger condition

$$\sum_{n=1}^{\infty}(1 - |z_n|)^{\alpha} < \infty$$

for all $\alpha > 1/2$ and yet $B' \notin \mathcal{N}$. We remind the reader that the Nevanlinna class $\mathcal{N}$ is so large that it contains all Hardy spaces H^p. Frostman's construction is such that every point of $\mathbb{T}$ is an accumulation point of the zeros of B. See Sect. 7.6. However, P. Ahern gave an example for which the set of accumulation points is much smaller. In fact, it can even be reduced to one point on $\mathbb{T}$.

To state Ahern's result we need a new definition. Let E be a closed subset of $\mathbb{T}$. Hence, its complement $\mathbb{T} \setminus E$, as an open subset of $\mathbb{T}$, is a countable union of open intervals I_n, $n \geq 1$. Denote the arc length of I_n by ℓ_n. We say that E is a *Carleson set* if $|E| = 0$ and

$$\sum_{n \geq 1} \ell_n \log 1/\ell_n < \infty.$$

Theorem 7.12 (Ahern [1]) *Let E be a closed subset of $\mathbb{T}$. Suppose that E has Lebesgue measure zero, and that it is not a Carleson set. Then there is a Blaschke sequence $(z_n)_{n \geq 1}$ which satisfies the following properties:*

(i) The set of accumulation points of $(z_n)_{n \geq 1}$ on $\mathbb{T}$ is a subset of E;
(ii) The sequence $(z_n)_{n \geq 1}$ satisfies

$$\sum_{n=1}^{\infty}(1 - |z_n|)^{\alpha} < \infty$$

for all $\alpha > 1/2$;
(iii) The Blaschke product B formed with $(z_n)_{n \geq 1}$ is such that $B' \notin \mathcal{N}$.

Proof. Without loss of generality, assume that $1 \in E$. Since E is closed, its complement is at most a countable union of open arcs, say

$$\mathbb{T} \setminus E = \bigcup_{n=1}^{\infty} I_n,$$

where $I_n = \{e^{i\theta} : 0 < \alpha_n < \theta < \beta_n < 2\pi\}$. For simplicity, let us write $\varepsilon_n = \beta_n - \alpha_n$. The assumption $|E| = 0$ is equivalent to

$$\sum_{n=1}^{\infty} \varepsilon_n = 2\pi \qquad (7.11)$$

and that E is not a Carleson set means that

$$\sum_{n=1}^{\infty} \varepsilon_n \, \log 1/\varepsilon_n = \infty.$$

Hence, we can choose a sequence $(\delta_n)_{n \geq 1}$ such that $0 < \delta_n < 1$ and

$$\lim_{n \to \infty} \delta_n = 0 \qquad (7.12)$$

but still

$$\sum_{n=1}^{\infty} \delta_n \, \varepsilon_n \, \log 1/\varepsilon_n = \infty. \qquad (7.13)$$

Then, for $n \geq 1$, put

$$\lambda_n = \left(1 - \varepsilon_n^{2-\delta_n}\right) e^{i\alpha_n} \qquad \text{and} \qquad \eta_n = \left(1 - \varepsilon_n^{2-\delta_n}\right) e^{i\beta_n}.$$

The union of $(\lambda_n)_{n \geq 1}$ and $(\eta_n)_{n \geq 1}$, with an appropriate indexing, is our Blaschke sequence $(z_n)_{n \geq 1}$.

According to the construction, no point of the Blaschke sequence is in the open sector

$$\{ re^{i\theta} : 0 < r < 1 \text{ and } \alpha_n < \theta < \beta_n \},$$

and thus the property (i) clearly holds. Therefore, by Theorem 1.5, the Blaschke product B has in fact an analytic extension across each I_n.

For any fixed $\alpha > 1/2$, by (7.12), we eventually have $(2 - \delta_n)\alpha \geq 1$, say for $n \geq N = N(\alpha)$, and thus

$$\sum_{n=N}^{\infty} (1 - |z_n|)^{\alpha} = 2 \sum_{n=N}^{\infty} \varepsilon_n^{(2-\delta_n)\alpha} \leq 2 \sum_{n=N}^{\infty} \varepsilon_n < \infty.$$

This establishes part (ii).

Finally, by (4.8), at least for each $e^{i\theta} \in \cup_{n=1}^{\infty} I_n$ we have

$$|B'(e^{i\theta})| = \sum_{n=1}^{\infty} \frac{1 - |\lambda_n|^2}{|e^{i\theta} - \lambda_n|^2} + \sum_{n=1}^{\infty} \frac{1 - |\eta_n|^2}{|e^{i\theta} - \eta_n|^2}.$$

We exploit this formula to obtain a lower estimation for $|B'|$. Write $\lambda_n = r_n e^{i\alpha_n}$, where $r_n = 1 - \varepsilon_n^{2-\delta_n}$. Then it is easy to verify that

$$|e^{i\theta} - \lambda_n|^2 = (1 - r_n)^2 + 4r_n \, \sin^2((\theta - \alpha_n)/2)$$
$$\leq (1 - r_n)^2 + (\theta - \alpha_n)^2.$$

Therefore, we have

$$|B'(e^{i\theta})| \geq \frac{1 - r_n}{(1 - r_n)^2 + (\theta - \alpha_n)^2}, \tag{7.14}$$

where $e^{i\theta}$ is any point of $\mathbb{T}$. In particular, the lower estimation

$$|B'(e^{i\theta})| \geq \frac{1 - r_n}{(1 - r_n)^2 + (\beta_n - \alpha_n)^2}$$

holds uniformly on I_n. Hence,

$$|B'(e^{i\theta})| \geq \frac{\varepsilon_n^{(2-\delta_n)}}{\varepsilon_n^{2(2-\delta_n)} + \varepsilon_n^2} = \frac{1}{\varepsilon_n^{\delta_n} + \varepsilon_n^{2-\delta_n}} \geq \frac{1}{2\varepsilon_n^{\delta_n}},$$

which implies

$$\int_{I_n} \log^+ |B'(e^{i\theta})| \, d\theta \geq (-\log 2)\,\varepsilon_n + \delta_n\,\varepsilon_n\,\log 1/\varepsilon_n.$$

Therefore, by (7.11) and (7.13),

$$\int_{\mathbb{T}} \log^+ |B'(e^{i\theta})| \, d\theta = \sum_{n=1}^{\infty} \int_{I_n} \log^+ |B'(e^{i\theta})| \, d\theta = \infty.$$

In other words, $B' \notin \mathcal{N}$, which is part (iii).

The success of the preceding construction is due to the fact that the sequence $(z_n)_{n\geq 1}$ tends *very* tangentially to the points of $\mathbb{T}$. This imprecise statement is somehow justified by the following result. This phenomenon will be further studied in Sect. 8.5.

Theorem 7.13 *If the zeros of a Blaschke product B all lie in a Stolz domain, then*

$$B' \in \bigcap_{0 < p < \frac{1}{2}} H^p(\mathbb{D}).$$

Remark. Corollary 8.17 is a generalization of this result.

Proof. (Girela–Peláez–Vukotić [24]) By (4.16),

$$|B'(z)| \leq \sum_{n=1}^{\infty} \frac{1 - |z_n|^2}{|1 - \bar{z}_n z|^2}, \qquad (z \in \mathbb{D}).$$

Suppose that all zeros of B are in the Stolz domain $S_C(e^{i\theta_0})$. Hence,

$$|B'(z)| \leq 4(2 + C)^2 \frac{\sum_{n=1}^{\infty}(1 - |z_n|^2)}{|e^{i\theta_0} - z|^2}, \qquad (z \in \mathbb{D}).$$

This is enough since

$$\frac{1}{(1-z)^2} \in \bigcap_{0<p<\frac{1}{2}} H^p(\mathbb{D}).$$

There is a Blaschke sequence B whose zeros are on the interval $[0,1)$ and yet $B' \notin H^{1/2}(\mathbb{D})$. See Examples 8.10 and 8.15. Hence, Theorem 7.13 is sharp.

7.5 A Sufficient Condition for the Existence of $B'(e^{i\theta})$

In this section, we study another essential theorem of Frostman, similar to Theorem 7.1, about the local boundary behavior of the derivative of a Blaschke product.

Theorem 7.14 (Frostman [23]) *Let* $(z_n)_{n\geq 1}$ *be a Blaschke sequence, and fix* $e^{i\theta} \in \mathbb{T}$. *Then the Blaschke product*

$$B(z) = \prod_{n=1}^{\infty} \frac{|z_n|}{z_n} \frac{z_n - z}{1 - \bar{z}_n z}, \qquad (z \in \mathbb{D}), \tag{7.15}$$

has angular derivative in the sense of Carathéodory at $e^{i\theta}$ *if and only if*

$$\sum_{n=1}^{\infty} \frac{1 - |z_n|}{|e^{i\theta} - z_n|^2} < \infty. \tag{7.16}$$

Moreover, in this case, we have

$$B'(e^{i\theta}) = e^{-i\theta} B(e^{i\theta}) \sum_{n=1}^{\infty} \frac{1 - |z_n|^2}{|e^{i\theta} - z_n|^2}.$$

Proof. Fix a Stolz domain $S_C(e^{i\theta})$. The assumption (7.16) a priori implies (7.2), and thus B and all its subproducts are continuous functions on $\overline{S_C}(e^{i\theta})$, the closure of $S_C(e^{i\theta})$ in the complex plane, and they have unimodular radial limits at the boundary point $e^{i\theta}$. By (4.2),

$$B'(z) = -\sum_{n=1}^{\infty} \frac{|z_n|}{z_n} \frac{1 - |z_n|^2}{(1 - \bar{z}_n z)^2} B_n(z), \qquad (z \in \mathbb{D}), \tag{7.17}$$

where

$$B_n(z) = \frac{B(z)}{\frac{|z_n|}{z_n} \frac{z_n - z}{1 - \bar{z}_n z}}. \tag{7.18}$$

Since

$$\left| \frac{|z_n|}{z_n} \frac{1 - |z_n|^2}{(1 - \bar{z}_n z)^2} B_n(z) \right| \leq 8(2 + C)^2 \frac{1 - |z_n|}{|e^{i\theta} - z_n|^2}, \qquad (z \in \overline{S_C(e^{i\theta})}),$$

we conclude that the series in (7.17) is uniformly convergent on $\overline{S_C(e^{i\theta})}$. Moreover, by Theorem 7.1, each term

$$\frac{|z_n|}{z_n} \frac{1 - |z_n|^2}{(1 - \bar{z}_n z)^2} B_n(z)$$

is a continuous function on $\overline{S_C(e^{i\theta})}$. Thus, by the Weierstrass M-test, B' is also continuous on $\overline{S_C(e^{i\theta})}$. Therefore,

$$B'(e^{i\theta}) = \lim_{\substack{z \longrightarrow e^{i\theta} \\ \vartriangleleft}} B'(z)$$

exists and, by (7.18),

$$B'(e^{i\theta}) = -\sum_{n=1}^{\infty} \frac{|z_n|}{z_n} \frac{1 - |z_n|^2}{(1 - \bar{z}_n e^{i\theta})^2} B_n(e^{i\theta})$$

$$= -\sum_{n=1}^{\infty} \frac{1 - |z_n|^2}{(1 - \bar{z}_n e^{i\theta})} \frac{B(e^{i\theta})}{z_n - e^{i\theta}}$$

$$= e^{-i\theta} B(e^{i\theta}) \sum_{n=1}^{\infty} \frac{1 - |z_n|^2}{|e^{i\theta} - z_n|^2}.$$

To prove that (7.16) is also necessary, we assume

$$\sum_{n=1}^{\infty} \frac{1 - |z_n|}{|e^{i\theta} - z_n|^2} = \infty$$

and then we show that B does not have an angular derivative in the sense of Carathéodory at $e^{i\theta}$. If the radial limit of B at $e^{i\theta}$ does not exist, or if it exists but is not unimodular, the claim is trivial. Hence, assume that the radial limit of B at $e^{i\theta}$ exists and $|B(e^{i\theta})| = 1$. Then

$$|B(re^{i\theta})|^2 = \prod_{n=1}^{\infty} \left(1 - \frac{(1 - |z_n|^2)(1 - r^2)}{|1 - \bar{z}_n re^{i\theta}|^2} \right)$$

$$\leq \exp \left\{ -\sum_{n=1}^{\infty} \frac{(1 - |z_n|^2)(1 - r^2)}{|1 - \bar{z}_n re^{i\theta}|^2} \right\}.$$

Thus, we must have

$$\lim_{r \to 1} \sum_{n=1}^{\infty} \frac{(1 - |z_n|^2)(1 - r^2)}{|1 - \bar{z}_n re^{i\theta}|^2} = 0.$$

Since the inequality $e^{-t} \leq 1 - t/2$ holds for small positive numbers, we obtain

$$|B(re^{i\theta})|^2 \leq 1 - \frac{1}{2} \sum_{n=1}^{\infty} \frac{(1 - |z_n|^2)(1 - r^2)}{|1 - \bar{z}_n re^{i\theta}|^2}.$$

Hence,

$$\frac{1 - |B(re^{i\theta})|^2}{1 - r^2} \geq \frac{1}{2} \sum_{n=1}^{\infty} \frac{1 - |z_n|^2}{|1 - \bar{z}_n re^{i\theta}|^2}.$$

Therefore, by Fatou's lemma,

$$\liminf_{r \to 1} \frac{1 - |B(re^{i\theta})|}{1 - r} \geq \frac{1}{4} \sum_{n=1}^{\infty} \frac{1 - |z_n|}{|1 - \bar{z}_n e^{i\theta}|^2} = \infty.$$

Theorem 4.8 now ensures that the derivative of B in the sense of Carathéodory at $e^{i\theta}$ does not exist.

7.6 The Global Behavior of B'

Theorem 7.1, which treats the local behavior of a Blaschke product, was applied to obtain global results like Theorems 7.4 and 7.8. In the same manner, we apply Theorem 7.14 to obtain global results about the behavior of B'.

Theorem 7.15 (Frostman [23]) *Let $0 < \alpha < 1/2$, let $(z_n)_{n \geq 1}$ be a Blaschke sequence satisfying the stronger condition*

$$\sum_{n=1}^{\infty} (1 - |z_n|)^{\alpha} < \infty,$$

and let B be the Blaschke product formed with this sequence. Then B has an angular derivative in the sense of Carathéodory at all points of $\mathbb{T}$ except possibly on a set of 2α-capacity zero.

Proof. Corresponding to each zero z_n, $n \geq 1$, consider the open interval I_n of length $(1 - |z_n|)^{1/2}$ on $\mathbb{T}$ whose middle point is on the ray passing through z_n. Let

$$V_n = \bigcup_{k=n}^{\infty} I_k, \qquad (n \geq 1),$$

and let

$$V = \bigcap_{n=1}^{\infty} V_n.$$

Hence, we have

$$\mu_{2\alpha}(V_n) \leq \sum_{k=n}^{\infty} (1 - |z_n|)^{\alpha},$$

which implies $\mu_{2\alpha}(V) = 0$. Therefore, we necessarily have $C_{2\alpha}(V) = 0$.

The open set V_n is defined such that, by the triangle inequality,

$$|e^{i\theta} - z_k| \geq \frac{1}{2} (1 - |z_n|)^{1/2} - (1 - |z_k|) \tag{7.19}$$

$$\geq \frac{1}{4} (1 - |z_n|)^{1/2}, \qquad (k \geq n), \tag{7.20}$$

for all $e^{i\theta} \notin V_n$. Let $E_n \subset (\mathbb{T} \setminus V_n)$ be the set of all points $e^{i\theta}$ such that

$$\sum_{k=1}^{\infty} \frac{1 - |z_k|}{|e^{i\theta} - z_k|^2} = \infty. \tag{7.21}$$

According to Theorem 7.14, at all points of the complement of $E_n \cup V_n$, B has angular derivative in the sense of Carathéodory. Hence, we proceed to show that $C_{2\alpha}(E_n) = 0$.

Suppose, on the contrary, that $C_{2\alpha}(E_n) > 0$. Thus, by Theorem 2.3, there is a positive Borel Measure μ, $\mu \not\equiv 0$, whose support is in E_n, and a positive constant C, such that

$$P_{\mu,2\alpha}(z) = \int_{\mathbb{C}} \frac{d\mu(e^{i\theta})}{|e^{i\theta} - z|^{2\alpha}} \leq C, \qquad (z \in \mathbb{C}).$$

Let

$$I = \int_{\mathbb{C}} \left(\sum_{k=n}^{\infty} \frac{1 - |z_k|}{|e^{i\theta} - z_k|^2} \right) d\mu(e^{i\theta}).$$

On the one hand, by (7.20), we have

$$I = \sum_{k=n}^{\infty} (1 - |z_k|)^{\alpha} \int_{\mathbb{C}} \left(\frac{1 - |z_k|}{|e^{i\theta} - z_k|^2} \right)^{1-\alpha} \frac{d\mu(e^{i\theta})}{|e^{i\theta} - z_n|^{2\alpha}}$$

$$\leq 16^{1-\alpha} \sum_{k=n}^{\infty} (1 - |z_k|)^{\alpha} \int_{\mathbb{C}} \frac{d\mu(e^{i\theta})}{|e^{i\theta} - z_k|^{2\alpha}}$$

$$= 16^{1-\alpha} \sum_{k=n}^{\infty} (1 - |z_k|)^{\alpha} P_{\mu,2\alpha}(z_k) \leq C \, 16^{1-\alpha} \sum_{n=1}^{\infty} (1 - |z_n|)^{\alpha} < \infty.$$

But, on the other hand, by (7.21),

$$I = \int_{E_n} \left(\sum_{k=n}^{\infty} \frac{1 - |z_k|}{|e^{i\theta} - z_k|^2} \right) d\mu(e^{i\theta}) = \infty,$$

which is a contradiction.

Let

$$E = V \cup \left(\bigcup_{n=1}^{\infty} E_n \right).$$

Clearly, $C_{2\alpha}(E) = 0$, and at each point of $\mathbb{T} \setminus E$, B has angular derivative in the sense of Carathéodory.

The preceding result is almost sharp in the following sense. Given a sequence of positive numbers $(r_n)_{n \geq 1}$ satisfying

$$\sum_{n=1}^{\infty} (1 - r_n) < \infty,$$

but, for some $0 < \alpha < 1/2$,

$$\sum_{n=1}^{\infty} (1 - r_n)^\alpha = \infty,$$

we are able to find a sequence of arguments $(\theta_n)_{n \geq 1}$ such that the Blaschke product formed with zeros $(r_n e^{i\theta_n})_{n \geq 1}$ does not have angular derivative in the sense of Carathéodory on a set of positive β–capacity, for all $\beta < 2\alpha$.

Let E be the set constructed on $\mathbb{T}$ with

$$\ell_n = 2^{-n/2\alpha} n^{-2/\alpha}$$

as described at the end of Sect. 2.3. The condition $2\ell_n < \ell_{n-1}$ is still satisfied. Hence, by (2.14), we have

$$\mu_\beta(E) = C \lim_{n \to \infty} 2^n \ell_n^\beta = C \lim_{n \to \infty} 2^{n(1 - \frac{\beta}{2\alpha})} n^{-2\beta/\alpha} = \infty$$

for each $\beta < 2\alpha$. Thus, by (2.11), we must also have

$$C_\beta(E) = \infty$$

for all $\beta < 2\alpha$. For simplicity, write $\varepsilon_n = 1 - r_n$. Then the relation

$$\sum_{n=1}^{\infty} \varepsilon_n^{2\alpha} \leq \sum_{n=1}^{\infty} 2^n \, \varepsilon_{2^n}^{2\alpha}$$

ensures that

$$\sum_{n=1}^{\infty} 2^n \, \varepsilon_{2^n}^{2\alpha} = \infty,$$

and therefore, for infinitely many values of n, we must have

$$2^n \, \varepsilon_{2^n}^{2\alpha} > \frac{1}{n^2}.$$

Equivalently, this means that, for infinitely many values of n,

$$\varepsilon_{2^n} > \ell_n. \tag{7.22}$$

As we did before, let θ_k be such that $z_k = r_k e^{i\theta_k}$ stays on the ray passing through the middle point of I_k. Fix $e^{i\theta} \in E$. Hence, $e^{i\theta} \in I_k$ for some k with $2^{n-1} \le k < 2^n$. Thus, by (7.22),

$$\frac{\varepsilon_k}{|e^{i\theta} - z_k|^2} \ge \frac{\varepsilon_k}{\varepsilon_k^2 + \frac{\ell_n^2}{4}}$$

$$\ge \frac{1}{\varepsilon_k + \frac{\ell_n^2}{4\varepsilon_{2^n}}} \ge \frac{4}{5}$$

for infinitely many values of n. Hence, the series (7.16) diverges and, by Theorem 7.14, we conclude that B does not have angular derivative in the sense of Carathéodory at any $e^{i\theta} \in E$.

Theorem 7.15 along with the preceding observation can be stated in the following technical language.

Corollary 7.16 *Let* $(z_n)_{n \ge 1}$ *be a Blaschke sequence, and let* B *be the Blaschke product formed with this sequence. Let* $E \subset \mathbb{T}$ *be the set of all points at which* B *does not have angular derivative in the sense of Carathéodory. If, for some* $0 < \alpha < 1/2$,

$$\sum_{n=1}^{\infty} (1 - |z_n|)^\alpha < \infty$$

then the Hausdorff dimension of E *is at most* 2α, *and if*

$$\sum_{n=1}^{\infty} (1 - |z_n|)^\alpha = \infty$$

then the Hausdorff dimension of E *can be at least* 2α.

We now treat the case $\alpha = 1/2$.

Theorem 7.17 (Frostman [23]) *Let* $(z_n)_{n \ge 1}$ *be a Blaschke sequence satisfying the stronger condition*

$$\sum_{n=1}^{\infty} (1 - |z_n|)^{1/2} < \infty,$$

and let B *be the Blaschke product formed with this sequence. Then* B *has an angular derivative in the sense of Carathéodory at all points of* $\mathbb{T}$ *except possibly on a set of Lebesgue measure zero.*

Proof. We apply the same construction exploited in the proof of Theorem 7.15. We need to show that E has the Lebesgue measure zero.

In the first place, according to the definition of I_k,

$$|V_n| \leq \sum_{k=n}^{\infty} |I_k| = \sum_{k=n}^{\infty} (1 - |z_n|)^{1/2},$$

and thus $|V| = \lim_{n \to \infty} |V_n| = 0$. Secondly, for n large enough,

$$\int_{\mathbb{T} \setminus V_n} \left(\sum_{k=n}^{\infty} \frac{1 - |z_k|}{|e^{i\theta} - z_k|^2} \right) d\theta \leq \sum_{k=n}^{\infty} \int_{\mathbb{T} \setminus I_k} \frac{1 - |z_k|}{|e^{i\theta} - z_k|^2} \, d\theta$$

$$\leq 2 \sum_{k=n}^{\infty} \int_{(1-|z_k|)^{1/2}}^{\pi} \frac{1 - |z_k|}{(1 - |z_k|)^2 + 2|z_k| \, t^2/\pi^2} \, dt$$

$$\leq 2\pi^2 \sum_{k=n}^{\infty} (1 - |z_k|)^{1/2},$$

which implies

$$\int_{E_n} \left(\sum_{k=1}^{\infty} \frac{1 - |z_k|}{|e^{i\theta} - z_k|^2} \right) d\theta \leq \int_{\mathbb{T} \setminus V_n} \left(\sum_{k=1}^{\infty} \frac{1 - |z_k|}{|e^{i\theta} - z_k|^2} \right) d\theta < \infty.$$

Hence, we must have $|E_n| = 0$, $n \geq 1$.

The preceding result is almost sharp in the following sense. Given a sequence of positive numbers $(r_n)_{n \geq 1}$ satisfying

$$\sum_{n=1}^{\infty} (1 - r_n) < \infty,$$

but

$$\sum_{n=1}^{\infty} (1 - r_n)^{1/2} = \infty,$$

we are able to find a sequence of arguments $(\theta_n)_{n \geq 1}$ such that the Blaschke product formed with zeros $(r_n e^{i\theta_n})_{n \geq 1}$ does not have angular derivative in the sense of Carathéodory at any point of $\mathbb{T}$. Put

$$\theta_n = \sum_{k=1}^{n} (1 - r_k)^{1/2}$$

and $z_n = r_n e^{i\theta_n}$, $n \geq 1$. Fix $e^{i\theta} \in \mathbb{T}$. Since

$$\theta_n \longrightarrow \infty \qquad \text{and} \qquad \theta_n - \theta_{n-1} \longrightarrow 0,$$

an $n \longrightarrow \infty$, there are infinitely many values of n such that $e^{i\theta}$ is on the arc $[e^{i\theta_{n-1}}, e^{i\theta_n}] \subset \mathbb{T}$. For each such n, we have

$$|e^{i\theta} - z_n|^2 \leq (1 - r_n)^2 + 4\sin^2(\frac{\theta_n - \theta}{2})$$
$$\leq (1 - r_n)^2 + (1 - r_n) \leq 2(1 - r_n).$$

Therefore, for infinitely many values of n, we have

$$\frac{1 - |z_n|}{|e^{i\theta} - z_n|^2} \geq \frac{1}{2}.$$

Thus,

$$\sum_{n=1}^{\infty} \frac{1 - |z_n|}{|e^{i\theta} - z_n|^2} = \infty$$

and, in the light of Theorem 7.14, we conclude that B does not have angular derivative in the sense of Carathéodory at any point of $\mathbb{T}$.

As a matter of fact, we can say a bit more about the Blaschke product we just constructed. According to Riesz's theorem, B has unimodular radial limits almost every where on $\mathbb{T}$. Hence, as we saw in the proof of Theorem 7.14, at any such point,

$$\liminf_{r \to 1} \frac{1 - |B(re^{i\theta})|}{1 - r} = \infty$$

which, by Theorem 4.8, implies

$$\lim_{r \to 1} |B'(re^{i\theta})| = \infty$$

for almost all $e^{i\theta} \in \mathbb{T}$. Therefore, certainly $B' \notin N$. See also Ahern's construction (Theorem 7.12).

Chapter 8
H^p-Means of B'

8.1 When Do We Have $B' \in H^1(\mathbb{D})$?

Theorem 5.4 provided a necessary and sufficient condition for the inclusion $B' \in H^p(\mathbb{D})$. This result, with some variation, is restated below . We will also see how the following lemma can be used to produce some families of Blaschke products that will be used throughout the text as illustrating examples.

Lemma 8.1 *Let B be a Blaschke product formed with the Blaschke sequence $z_n = r_n e^{i\theta_n}$, $n \geq 1$, and let $0 < p \leq \infty$. Then the following are equivalent:*

(i)
$$B' \in H^p(\mathbb{D});$$

(ii)
$$|B'(e^{i\theta})| = \sum_n \frac{1 - |z_n|^2}{|e^{i\theta} - z_n|^2} \in L^p(\mathbb{T});$$

(iii)
$$f_B(\theta) = \sum_{n=1}^{\infty} \frac{1 - r_n}{(1 - r_n)^2 + (\theta - \theta_n)^2} \in L^p.$$

Proof. This is a special case of Theorem 5.4. It is enough just to observe that $(1 - |z_n|^2) \asymp 1 - r_n$ and

$$|e^{i\theta} - z_n|^2 \asymp (1 - r_n)^2 + (\theta - \theta_n)^2. \tag{8.1}$$

While there is no ambiguity in saying

$$\sum_n \frac{1 - |z_n|^2}{|e^{i\theta} - z_n|^2} \in L^p(\mathbb{T}),$$

there is a little doubt about the choice of θ and θ_n in the definition of f_B. That is why indeed, in the statement of Lemma 8.1, we wrote $f_B \in L^p$ and did not mention which L^p should be considered. In fact, our choice should be such that (8.1) remains valid, and for this, we need $|\theta - \theta_n|$ to be small if $|e^{i\theta} - z_n|$ is small. A remedy is to write $B = B_1 B_2$, where B_1 and B_2 are the subproducts formed respectively with zeros in the right hand semidisc and left hand semidisc. Then, for f_{B_1}, we can take $-\pi/2 \leq \theta_n \leq \pi/2$ and $-\pi < \theta < \pi$. Now, we can say $B_1' \in H^p(\mathbb{D})$ if and only if $f_{B_1} \in L^p(-\pi, \pi)$. The treatment of B_2 is similar.

Corollary 8.2 *Let B be a Blaschke product. Then $B' \in H^1(\mathbb{D})$ if and only if B is a finite Blaschke product.*

Proof. By Lemma 8.1, $B' \in H^1(\mathbb{D})$ if and only if

$$\sum_n \frac{1 - |z_n|^2}{|e^{i\theta} - z_n|^2} \in L^1(\mathbb{T}).$$

But,

$$\frac{1}{2\pi} \int_0^{2\pi} \left(\sum_n \frac{1 - |z_n|^2}{|e^{i\theta} - z_n|^2} \right) d\theta = \sum_n \frac{1}{2\pi} \int_0^{2\pi} \frac{1 - |z_n|^2}{|e^{i\theta} - z_n|^2} d\theta = \sum_n 1.$$

Clearly, the sum is finite if and only if the index n runs through a finite set.

Corollary 8.2 reveals that for infinite Blaschke products, we can hope for the relation $B' \in H^p(\mathbb{D})$ only if $0 < p < 1$. In Sect. 7.4, we showed that this hope cannot be achieved if we do not put some stronger restrictions on the rate of growth of zeros of B. In other words, there are infinite Blaschke products whose derivatives are not in any Hardy space. However, in Sect. 8.2, we present sufficient conditions which guarantee $B' \in H^p(\mathbb{D})$ for some $0 < p < 1$. In particular, there is an infinite Blaschke product B such that

$$B' \in \bigcap_{0 < p < 1} H^p(\mathbb{D}).$$

See Example 8.5.

The impossibility of $B' \in H^1(\mathbb{D})$ for infinite Blaschke products can also be looked at from a different angle. Since B has infinitely many zeros that cluster on some points of $\mathbb{T}$ and at the same time, by Theorem 1.7, B is unimodular almost everywhere on $\mathbb{T}$, then B is never continuous on $\overline{\mathbb{D}}$. Therefore, the following result is another manifestation that for no infinite Blaschke product B, the relation $B' \in H^1(\mathbb{D})$ holds.

Theorem 8.3 (Privalov [37, 38]) *Let f be analytic on $\mathbb{D}$. Then $f' \in H^1(\mathbb{D})$ if and only if f is continuous on $\overline{\mathbb{D}}$ and absolutely continuous on $\mathbb{T}$.*

Proof. Suppose that $f' \in H^1(\mathbb{D})$. Hence, the limit

$$\phi(t) = \lim_{r \to 1} f'(re^{it})$$

exists for almost all $t \in [0, 2\pi]$ and $\phi \in L^1([0, 2\pi])$. Moreover, $ie^{it}\phi(t)$ is the boundary values of the H^1-function $izf'(z)$, which is vanishing at the origin, and thus we must have

$$\int_0^{2\pi} ie^{it}\phi(t)\, dt = 0. \tag{8.2}$$

Hence, the representation

$$ire^{i\theta} f'(re^{i\theta}) = \frac{1}{2\pi} \int_0^{2\pi} \frac{1 - r^2}{1 + r^2 - 2r\cos(\theta - t)} ie^{it}\phi(t)\, dt \tag{8.3}$$

holds for all $re^{i\theta} \in \mathbb{D}$.

Define

$$\Phi(t) = \int_0^t ie^{is}\phi(s)\, ds, \qquad (t \in [0, 2\pi]).$$

Since $\phi \in L^1([0, 2\pi])$, the function Φ is *absolutely continuous* on $[0, 2\pi]$ and, by (8.2), $\Phi(0) = \Phi(2\pi) = 0$. Thus, we may interpret Φ as an absolutely continuous function on $\mathbb{T}$.

The left side of (8.3) is equal to

$$\frac{\partial}{\partial \theta}\left(f(re^{i\theta}) \right),$$

and, for the right side, by doing the integration by parts, we obtain

$$\frac{\partial}{\partial \theta}\left(f(re^{i\theta}) \right) = \frac{1}{2\pi} \int_0^{2\pi} \frac{1 - r^2}{1 + r^2 - 2r\cos(\theta - t)} \Phi'(t)\, dt$$

$$= -\frac{1}{2\pi} \int_0^{2\pi} \frac{\partial}{\partial t}\left(\frac{1 - r^2}{1 + r^2 - 2r\cos(\theta - t)} \right) \Phi(t)\, dt$$

$$= \frac{1}{2\pi} \int_0^{2\pi} \frac{(1 - r^2)(-2r\sin(\theta - t))}{(1 + r^2 - 2r\cos(\theta - t))^2} \Phi(t)\, dt$$

$$= \frac{1}{2\pi} \int_0^{2\pi} \frac{\partial}{\partial \theta}\left(\frac{1 - r^2}{1 + r^2 - 2r\cos(\theta - t)} \right) \Phi(t)\, dt$$

$$= \frac{\partial}{\partial \theta}\left(\frac{1}{2\pi} \int_0^{2\pi} \frac{1 - r^2}{1 + r^2 - 2r\cos(\theta - t)} \Phi(t)\, dt \right).$$

The essential part of proof is the second identity above, which works since Φ is absolutely continuous on $\mathbb{T}$. Then the equation

$$\frac{\partial}{\partial \theta}\left(f(re^{i\theta}) \right) = \frac{\partial}{\partial \theta}\left(\frac{1}{2\pi} \int_0^{2\pi} \frac{1 - r^2}{1 + r^2 - 2r\cos(\theta - t)} \Phi(t)\, dt \right)$$

implies

$$f(re^{i\theta}) = \frac{1}{2\pi} \int_0^{2\pi} \frac{1-r^2}{1+r^2-2r\cos(\theta-t)}\, \Phi(t)\, dt + C(r),$$

where $C(r)$ is constant with respect to θ. But, $C(r)$ is the difference of two complex valued harmonic functions. Hence, we must have

$$C(r) = a\log r + b,$$

where a and b are arbitrary constants. However, the continuity at the origin forces $a = 0$. In short, we have the representation

$$f(re^{i\theta}) = \frac{1}{2\pi} \int_0^{2\pi} \frac{1-r^2}{1+r^2-2r\cos(\theta-t)}\, \big(\Phi(t)+b\big)\, dt, \qquad (re^{i\theta} \in \mathbb{D}),$$

and this representation reveals that $\Phi + b$ is actually the boundary trace of f on $\mathbb{T}$. Hence, f is continuous on $\overline{\mathbb{D}}$ and f is absolutely continuous on $\mathbb{T}$.

Now, suppose that f is analytic on $\mathbb{D}$, continuous on $\overline{\mathbb{D}}$ and absolutely continuous on $\mathbb{T}$. Then we have the representation

$$f(re^{i\theta}) = \frac{1}{2\pi} \int_0^{2\pi} \frac{1-r^2}{1+r^2-2r\cos(\theta-t)}\, f(e^{it})\, dt, \qquad (re^{i\theta} \in \mathbb{D}).$$

Differentiating with respect to θ gives

$$ire^{i\theta} f'(re^{i\theta}) = \frac{1}{2\pi} \int_0^{2\pi} \frac{\partial}{\partial \theta}\left(\frac{1-r^2}{1+r^2-2r\cos(\theta-t)} \right) f(e^{it})\, dt$$

$$= \frac{1}{2\pi} \int_0^{2\pi} \frac{(1-r^2)(-2r\sin(\theta-t))}{\big(1+r^2-2r\cos(\theta-t)\big)^2} f(e^{it})\, dt$$

$$= -\frac{1}{2\pi} \int_0^{2\pi} \frac{\partial}{\partial t}\left(\frac{1-r^2}{1+r^2-2r\cos(\theta-t)} \right) f(e^{it})\, dt.$$

Since f is absolutely continuous on $\mathbb{T}$, we can integrate by parts to obtain

$$ire^{i\theta} f'(re^{i\theta}) = \frac{1}{2\pi} \int_0^{2\pi} \frac{1-r^2}{1+r^2-2r\cos(\theta-t)} \left(\frac{d}{dt} f(e^{it}) \right) dt.$$

This was a crucial step. The absolute continuity of f also means that

$$\frac{d}{dt} f(e^{it}) \in L^1(\mathbb{T}).$$

Hence, $izf'(z) \in H^1(\mathbb{T})$, which implies $f' \in H^1(\mathbb{T})$.

Note that, under the assumptions of the preceding theorem, we incidentally proved

$$that f(e^{i\theta}) = f(1) + \int_0^\theta ie^{it} f'(e^{it})\, dt, \qquad (e^{i\theta} \in \mathbb{T}),$$

where

$$f'(e^{it}) = \lim_{r \to 1} f'(re^{it}), \qquad (a.e.\ e^{it} \in \mathbb{T}),$$

and

$$f(e^{it}) = \lim_{r \to 1} f(re^{it}), \qquad (e^{it} \in \mathbb{T}).$$

This fact may be also rewritten as

$$\frac{d}{d\theta} f(e^{i\theta}) = ie^{i\theta} f'(e^{i\theta}), \qquad (a.e.\ e^{i\theta} \in \mathbb{T}),$$

which looks like the chain rule for derivation.

8.2 A Sufficient Condition for $B' \in H^p(\mathbb{D})$, $0 < p < 1$

Theorem 7.12 shows that the condition

$$\sum_{n=1}^{\infty} (1 - |z_n|)^{\alpha} < \infty$$

for some (or even all) $\alpha > 1/2$ is not enough to deduce $B' \in H^p(\mathbb{D})$. Hence, a more restrictive condition is needed. The first result of this type was obtained by D. Protas.

Theorem 8.4 (Protas [39]) *Let $(z_n)_{n \geq 1}$ be a Blaschke sequence satisfying the stronger condition*

$$\sum_{n=1}^{\infty} (1 - |z_n|)^{\alpha} < \infty$$

for some $0 < \alpha < 1/2$. Then $B' \in H^{1-\alpha}(\mathbb{D})$.

Proof. Write $B = \prod_{n=1}^{\infty} b_{z_n}$. By (4.23),

$$\|b'_{z_n}\|_{1-\alpha}^{1-\alpha} \leq C_\alpha (1 - |z_n|)^\alpha.$$

Hence, $b'_{z_n} \in H^{1-\alpha}(\mathbb{D})$ and, by assumption,

$$\sum_{n=1}^{\infty} \|b'_{z_n}\|_{1-\alpha}^{1-\alpha} < \infty.$$

Therefore, by Theorem 5.5, $B' \in H^{1-\alpha}(\mathbb{D})$.

Fix $0 < \alpha < 1/2$. Then there is a Blaschke product with zeros $(r_n e^{i\theta_n})_{n \geq 1}$ such that

$$\sum_{n=1}^{\infty} (1 - r_n)^\alpha < \infty,$$

and thus $B' \in H^{1-\alpha}(\mathbb{D})$, but

$$B' \notin H^p(\mathbb{D})$$

for any $p > 1 - \alpha$. See Example 8.12. Hence, Theorem 8.4 is somehow sharp. However, its converse is not true. There is a Blaschke product B such that $B' \in H^{1-\alpha}(\mathbb{D})$, and yet

$$\sum_{n=1}^{\infty} (1 - r_n)^{\alpha} = \infty.$$

Example 8.5 Let $(r_n)_{n \geq 1}$ be any sequence in $[0, 1)$ such that

$$\sum_{n=1}^{\infty} (1 - r_n)^{\alpha} < \infty$$

for all $\alpha > 0$. A particular choice is $r_n = 1 - 2^{-n}$, $n \geq 1$. Let $(\theta_n)_{n \geq 1}$ be any arbitrary sequence of real numbers and put $z_n = r_n e^{i\theta_n}$, $n \geq 1$. Then, by Theorem 8.4, the Blaschke product B formed with $(z_n)_{n \geq 1}$ satisfies

$$B' \in \bigcap_{0 < p < 1} H^p(\mathbb{D}).$$

The condition

$$\sum_{n=1}^{\infty} (1 - |z_n|)^{\frac{1}{2}} < \infty$$

is not enough to ensure that $B' \in H^{\frac{1}{2}}(\mathbb{D})$. Hence, we need again a slightly more restrictive condition.

Theorem 8.6 (Protas [39]) *Let $(z_n)_{n \geq 1}$ be a Blaschke sequence satisfying the stronger condition*

$$\sum_{n=1}^{\infty} (1 - |z_n|)^{\frac{1}{2}} \log \frac{1}{(1 - |z_n|)} < \infty.$$

Then $B' \in H^{\frac{1}{2}}(\mathbb{D})$.

Proof. The proof is similar to the one given above for Theorem 8.4. First, By (4.23),

$$\|b'_{z_n}\|_{\frac{1}{2}}^{\frac{1}{2}} \leq C(1 - |z_n|)^{\frac{1}{2}} \log \frac{1}{(1 - |z_n|)}.$$

Hence, $b'_{z_n} \in H^{\frac{1}{2}}(\mathbb{D})$ and, by assumption,

$$\sum_{n=1}^{\infty} \|b'_{z_n}\|_{\frac{1}{2}}^{\frac{1}{2}} < \infty.$$

Theorem 5.5 now ensures that $B' \in H^{\frac{1}{2}}(\mathbb{D})$.

There is a Blaschke product with zeros $(r_n e^{i\theta_n})_{n \geq 1}$ such that

$$\sum_{n=1}^{\infty} (1 - |z_n|)^{\frac{1}{2}} \log \frac{1}{(1 - |z_n|)} < \infty,$$

and thus $B' \in H^{\frac{1}{2}}(\mathbb{D})$, but

$$B' \notin H^p(\mathbb{D})$$

for any $p > \frac{1}{2}$. Hence, in this regard, Theorem 8.6 is sharp. However, its converse is not true. This means that there is a Blaschke product B such that $B' \in H^{\frac{1}{2}}(\mathbb{D})$, and yet

$$\sum_{n=1}^{\infty} (1 - |z_n|)^{\frac{1}{2}} \log \frac{1}{(1 - |z_n|)} = \infty.$$

8.3 What Does $B' \in H^p(\mathbb{D})$, $0 < p < 1$, Imply?

In Sect. 8.2, we saw that a strong restriction on the rate of growth of $|z_n|$ forces B' to be in a Hardy space $H^p(\mathbb{D})$. None of those results were reversible. Nevertheless, a partial converse holds. We study them in this section. But, we emphasize that the following results are not reversible either.

Theorem 8.7 (Ahern–Clark [2]) *Let B be a Blaschke product formed with the Blaschke sequence $(z_n)_{n \geq 1}$. Let $1/2 < p < 1$, and suppose that $B' \in H^p(\mathbb{D})$. Then*

$$\sum_{n=1}^{\infty} (1 - |z_n|)^{\frac{1-p}{p}} < \infty.$$

Proof. By (4.23),

$$\|b'_{z_n}\|_p \geq C_p (1 - |z_n|)^{(1-p)/p},$$

and, by Theorem 5.5,

$$C_p \sum_{n=1}^{\infty} (1 - |z_n|)^{\frac{1-p}{p}} \leq \sum_{n=1}^{\infty} \|b'_{z_n}\|_p \leq \|B'\|_p < \infty.$$

Fix $1/2 < p < 1$. Then there is a Blaschke product B, even with zeros on the interval $(0, 1)$, such that $B' \in H^p(\mathbb{D})$, and thus

$$\sum_{n=1}^{\infty}(1-|z_n|)^{\frac{1-p}{p}} < \infty,$$

but

$$\sum_{n=1}^{\infty}(1-|z_n|)^{\frac{1-p}{p}-\varepsilon} = \infty,$$

for any $\varepsilon > 0$. See Example 8.19. Hence, Theorem 8.7 is sharp. However, its converse is not true. There is a Blaschke product B, even with zeros on the interval $(0,1)$, such that

$$\sum_{n=1}^{\infty}(1-|z_n|)^{\frac{1-p}{p}} < \infty,$$

and yet $B' \notin H^p(\mathbb{D})$. See Example 8.18.

To be able to better compare Theorems 8.4 and 8.7, we can write them in the following ways. Let $0 < \alpha < \frac{1}{2}$, and let $\frac{1}{2} < p < 1$. Then

$$\sum_{n=1}^{\infty}(1-|z_n|)^{\alpha} < \infty \Longrightarrow B' \in H^{1-\alpha}(\mathbb{D}),$$

$$\sum_{n=1}^{\infty}(1-|z_n|)^{\frac{\alpha}{1-\alpha}} = \infty \Longrightarrow B' \notin H^{1-\alpha}(\mathbb{D}),$$

or equivalently

$$B' \notin H^p(\mathbb{D}) \Longrightarrow \sum_{n=1}^{\infty}(1-|z_n|)^{1-p} = \infty,$$

$$B' \in H^p(\mathbb{D}) \Longrightarrow \sum_{n=1}^{\infty}(1-|z_n|)^{\frac{1-p}{p}} < \infty.$$

Clearly, none of the above implications is reversible.

Even though Theorems 8.4 and 8.7 together do not provide a necessary and sufficient condition for $B' \in H^p(\mathbb{D})$ in terms of a growth restriction on $|z_n|$, $n \geq 1$, the story is different if we consider the interpolating Blaschke products.

Theorem 8.8 (Cohn [14]) *Fix $0 < \alpha < \frac{1}{2}$. Let $(z_n)_{n\geq 1}$ be an interpolating Blaschke sequence, and let B be the corresponding Blaschke product. Then $B' \in H^{1-\alpha}(\mathbb{D})$ if and only if*

$$\sum_{n=1}^{\infty}(1-|z_n|)^{\alpha} < \infty.$$

Remark: The *only if* part works for $0 < \alpha < 1$.

Proof. The *if* part is the content of Protas theorem (Theorem 8.4). For the *only if* part, the assumption $B' \in H^{1-\alpha}(\mathbb{D})$ implies

$$\sum_{n=1}^{\infty} |B'(z_n)|^{1-\alpha} \, (1 - |z_n|) \lesssim \|B'\|_{1-\alpha}^{1-\alpha}.$$

But, by (4.5),

$$|B'(z_n)| \asymp \frac{1}{1 - |z_n|}.$$

Hence, the result follows.

Theorem 8.9 (Ahern–Clark [2]) *Let B be a Blaschke product formed with the Blaschke sequence $(z_n)_{n \geq 1}$. Suppose that $B' \in H^{1/2}(\mathbb{D})$. Then*

$$\sum_{n=1}^{\infty} (1 - |z_n|) \, \log^2 \frac{1}{(1 - |z_n|)} < \infty.$$

Proof. By (4.23),

$$\|b'_{z_n}\|_{1/2} \geq C(1 - |z_n|) \, \log^2 1/(1 - |z_n|),$$

and, by Theorem 5.5,

$$C \sum_{n=1}^{\infty} (1 - |z_n|) \, \log^2 1/(1 - |z_n|) \leq \sum_{n=1}^{\infty} \|b'_{z_n}\|_{1/2} \leq \|B'\|_{1/2} < \infty.$$

There is a Blaschke product B such that $B' \in H^{1/2}(\mathbb{D})$, and thus

$$\sum_{n=1}^{\infty} (1 - |z_n|) \, \log^2 \frac{1}{(1 - |z_n|)} < \infty,$$

but

$$\sum_{n=1}^{\infty} (1 - |z_n|) \, \log^{2+\varepsilon} \frac{1}{(1 - |z_n|)} = \infty$$

for all $\varepsilon > 0$. Hence, Theorem 8.9 is sharp. However, its converse is not true. There is a Blaschke product B such that

$$\sum_{n=1}^{\infty} (1 - |z_n|) \, \log^2 \frac{1}{(1 - |z_n|)} < \infty.$$

and yet $B' \notin H^{1/2}(\mathbb{D})$.

Example 8.10   The sequence $z_n = 1 - \frac{1}{n \log^2 n}$, $n \geq 2$, satisfies

$$(1 - |z_n|) \log^2 \frac{1}{(1 - |z_n|)} \asymp \frac{1}{n},$$

and thus, by Theorem 8.9, it gives a Blaschke product B with radial zeros such that $B' \notin H^{\frac{1}{2}}(\mathbb{D})$.

To better compare Theorems 8.6 and 8.9, we can write them in the following ways:

$$\sum_{n=1}^{\infty} (1 - |z_n|)^{\frac{1}{2}} \log \frac{1}{(1 - |z_n|)} < \infty \Longrightarrow B' \in H^{1/2}(\mathbb{D}),$$

$$\sum_{n=1}^{\infty} (1 - |z_n|) \log^2 \frac{1}{(1 - |z_n|)} = \infty \Longrightarrow B' \notin H^{1/2}(\mathbb{D}),$$

or equivalently

$$B' \notin H^{1/2}(\mathbb{D}) \Longrightarrow \sum_{n=1}^{\infty} (1 - |z_n|)^{\frac{1}{2}} \log \frac{1}{(1 - |z_n|)} = \infty,$$

$$B' \in H^{1/2}(\mathbb{D}) \Longrightarrow \sum_{n=1}^{\infty} (1 - |z_n|) \log^2 \frac{1}{(1 - |z_n|)} < \infty.$$

Clearly, none of these implications is reversible.

8.4 Some Examples of Blaschke Products

In this section, we give some specific examples of Blaschke products which were either promised before, or will be exploited later on. The family given below is a generalization of the Ahern–Clark example [2], which is reproduced after the lemma.

Lemma 8.11 *Let $\psi : (0,1) \longrightarrow (0, \infty)$ be a decreasing function such that*

$$\int_0^1 \psi(t) \, dt = \infty.$$

Let $0 < \alpha < 1$. Suppose that $(r_n e^{i\theta_n})_{n \geq 1}$ is a Blaschke sequence satisfying the following properties:

(i) $(\theta_n)_{n\geq 1}$ is a positive decreasing sequence which tends to zero;

(ii) $1 - r_n \asymp \theta_n - \theta_{n+1} \asymp (\psi(\theta_{n+1}))^{\frac{-1}{1-\alpha}}$.

Let B be the Blaschke product formed with the sequence $(r_n e^{i\theta_n})_{n\geq 1}$. Then

$$B' \notin H^{1-\alpha}(\mathbb{D}).$$

Proof. The assumptions are designed to obtain

$$f_B(\theta) = \sum_{k=1}^{\infty} \frac{1-r_k}{(1-r_k)^2 + (\theta-\theta_k)^2} \geq C \, (\psi(\theta))^{\frac{1}{1-\alpha}}$$

for $0 < \theta < \pi/2$, which, by Lemma 8.1, ensures that $B' \notin H^{1-\alpha}(\mathbb{D})$. To verify the last inequality, assume that $0 < \theta_{n+1} < \theta < \theta_n < \pi/2$. Then

$$
\begin{aligned}
f_B(\theta) &= \sum_{k=1}^{\infty} \frac{1-r_k}{(1-r_k)^2 + (\theta-\theta_k)^2} \\
&\geq \frac{1-r_n}{(1-r_n)^2 + (\theta-\theta_n)^2} \\
&\asymp \frac{1-r_n}{(1-r_n)^2 + (\theta_{n+1}-\theta_n)^2} \\
&\asymp (\psi(\theta_{n+1}))^{\frac{1}{1-\alpha}} \geq (\psi(\theta))^{\frac{1}{1-\alpha}}.
\end{aligned}
$$

Example 8.12 An specific example for Lemma 8.11 is

$$r_n = 1 - n^{-\frac{1}{\alpha}} \qquad \text{and} \qquad \theta_n = n^{1-\frac{1}{\alpha}},$$

where $0 < \alpha < 1$, and $\psi(t) = 1/t$. Then

$$
\begin{aligned}
\theta_n - \theta_{n+1} &= n^{1-\frac{1}{\alpha}} - (n+1)^{1-\frac{1}{\alpha}} \\
&= \frac{(1+\frac{1}{n})^{\frac{1}{\alpha}-1} - 1}{(n+1)^{\frac{1}{\alpha}-1}} \\
&\asymp \frac{1}{n^\alpha} \asymp \theta_{n+1}^{\frac{1}{1-\alpha}}.
\end{aligned}
$$

Other conditions of Lemma 8.11 are easily verified. Hence, $B' \notin H^{1-\alpha}(\mathbb{D})$. Note that for this conclusion, we cannot apply Theorem 8.7. It is also interesting to observe that if $0 < \alpha < 1/2$, then, for each $\alpha < \beta < 1/2$, we have

$$\sum_{n=1}^{\infty} (1-r_n)^\beta = \sum_{n=1}^{\infty} \frac{1}{n^{\beta/\alpha}} < \infty,$$

and thus, by Theorem 8.4, $B' \in H^{1-\beta}(\mathbb{D})$. Hence, we have a Blaschke product B such that

$$B' \in \bigcap_{0 < p < 1-\alpha} H^p(\mathbb{D}),$$

but

$$B' \notin H^{1-\alpha}(\mathbb{D}).$$

Example 8.13   While the last example sheds some light on the sharpness of Theorem 8.4, we can still work on the example and come up with a more delicate Blaschke product. Fix $0 < \alpha < 1/2$. Our goal is to produce a Blaschke sequence $(r_n e^{i\theta_n})_{n \geq 1}$ such that

$$\sum_{n=1}^{\infty} (1 - r_n)^\alpha < \infty,$$

and thus, by Theorem 8.4, $B' \in H^{1-\alpha}(\mathbb{D})$, but

$$B' \notin H^p(\mathbb{D}), \qquad (p > 1 - \alpha).$$

This example would be another manifestation of the sharpness of Theorem 8.4. To do so, pick any sequence

$$0 < \alpha_1 < \alpha_2 < \alpha_3 < \cdots < \alpha$$

with $\lim_{j \to \infty} \alpha_j = \alpha$. Then, as explained above, the Blaschke product B_j formed with the sequence

$$r_{n,j} = 1 - n^{-\frac{1}{\alpha_j}} \qquad \text{and} \qquad \theta_{n,j} = n^{1 - \frac{1}{\alpha_j}}$$

is such that $B_j' \notin H^{1-\alpha_j}(\mathbb{D})$, and yet

$$\sum_{n=1}^{\infty} (1 - r_{n,j})^\alpha < \infty.$$

To overcome a convergence difficulty, we delete a finite number of zeros such that

$$\sum_{n=N_j}^{\infty} (1 - r_{n,j})^\alpha \leq \frac{1}{2^j}.$$

Now, the Blaschke product $B = \prod_{j=1}^{\infty} B_j$ is well-defined and its zeros satisfy

$$\sum_{n=1}^{\infty} (1 - r_n)^\alpha = \sum_{j=1}^{\infty} \sum_{n=N_j}^{\infty} (1 - r_{n,j})^\alpha < \infty.$$

Moreover, since for any $p > 1 - \alpha$ there are (infinitely many) integers j such that $p > 1 - \alpha_j > 1 - \alpha$, we have

$$\|B'_j\|_p \geq \|B'_j\|_{1-\alpha_j} = \infty.$$

Hence, by Theorem 5.5,

$$\|B'\|_p \geq \|B'_j\|_p = \infty.$$

Lemma 8.14 *Let* $\psi : (0,1) \longrightarrow (0, \infty)$ *be a decreasing function such that*

$$\int_0^1 \psi(t) \, dt = \infty.$$

Let $(\varepsilon_n)_{n \geq 1}$ *be a sequence of positive decreasing numbers such that*

$$\sum_{n=1}^{\infty} \varepsilon_n < \infty$$

and

$$\sum_{n=1}^{N} \frac{1}{\varepsilon_n} \geq \psi^2(\varepsilon_{N+1}), \qquad (N \geq 1).$$

Let B *be the Blaschke product formed with the sequence* $z_n = 1 - \varepsilon_n$, $n \geq 1$. *Then*

$$B' \notin H^{1/2}(\mathbb{D}).$$

Remark: As the following proof shows, the values of ψ in a right neighborhood of zero are important. Hence, we may assume that ψ is defined on $(0, \delta)$ and $\int_0^{\delta} \psi(t) \, dt = \infty$. In the same spirit, the estimation $\sum_{n=1}^{N} 1/\varepsilon_n \geq \psi^2(\varepsilon_{N+1})$ is enough to be valid only for large values of N. Moreover, in the light of Theorem 8.9, we obtain a new family if $\sum_{n=1}^{\infty} (1 - |z_n|) \log^2 \frac{1}{(1-|z_n|)} < \infty$.

Proof. The assumptions are adjusted to obtain

$$f_B(t) = \sum_{n=1}^{\infty} \frac{\varepsilon_n}{\varepsilon_n^2 + t^2} \geq C \, \psi^2(t)$$

for $0 < t < 1$, which, by Lemma 8.1, ensures that $B' \notin H^{1/2}(\mathbb{D})$. To verify the last inequality, assume that $0 < \varepsilon_{N+1} < t < \varepsilon_N < 1$. Then

$$f_B(t) = \sum_{n=1}^{\infty} \frac{\varepsilon_n}{\varepsilon_n^2 + t^2} \geq \sum_{n=1}^{N} \frac{\varepsilon_n}{\varepsilon_n^2 + t^2}$$

$$\geq \sum_{n=1}^{N} \frac{1}{2\varepsilon_n} \geq \frac{1}{2} \psi^2(\varepsilon_{N+1}) \geq \frac{1}{2} \psi^2(t).$$

Example 8.15   A realization of lemma is the sequence $\varepsilon_n = 1/n \log^4 n$, $n \geq 2$, for which we should take

$$\psi(t) = \frac{1}{t \log^2 1/t}.$$

Then

$$\sum_{n=2}^{N} \frac{1}{\varepsilon_n} = \sum_{n=2}^{N} n \log^4 n \asymp \int_2^N t \log^4 t \, dt \asymp N^2 \log^4 N \asymp \psi^2(\varepsilon_{N+1}).$$

Hence, the corresponding Blaschke product has radial zeros and yet $B' \notin H^{1/2}(\mathbb{D})$. Note that for this example we cannot apply Theorem 8.9.

8.5 The Study of $\|B'\|_p$ When Zeros Are in a Stolz Domain

In Theorem 7.13, we showed that

$$B' \in \bigcap_{0 < p < \frac{1}{2}} H^p(\mathbb{D})$$

when the zeros of B all lie in a Stolz domain. Moreover, Examples 8.10 and 8.15 reveal that this result is sharp. In this section we study further this class of Blaschke products. If we put a growth restriction on $|z_n|$ and also assume that the zeros are in a Stolz domain, naturally we expect to obtain a better result, which is not promised by Theorem 8.4. The following theorem confirms this expectation.

Theorem 8.16 (Ahern–Clark [2]) *Let $(z_n)_{n \geq 1}$ be a Blaschke sequence satisfying the condition*

$$\sum_{n=1}^{\infty} (1 - |z_n|)^\alpha < \infty$$

for some $0 < \alpha \leq 1$. Suppose, moreover, that the zeros are in the generalized Stolz domain $S_{C,\delta}(e^{i\theta_0})$, for some fixed $C \geq 1$ and $0 < \delta \leq 1$. Then

$$B' \in \bigcap_{0 < p < \frac{\delta}{2 - \delta(1-\alpha)}} H^p(\mathbb{D}).$$

Proof. Without loss of generality, we may assume that $\theta_0 = 0$. According to Theorem 4.15,

$$|B'(e^{i\theta})| = \sum_{n=1}^{\infty} \frac{1 - |z_n|^2}{|e^{i\theta} - z_n|^2}, \qquad (e^{i\theta} \in \mathbb{T}).$$

But, on a generalized Stolz domain, we have

$$\frac{|e^{i\theta} - z_n|^\delta}{|e^{i\theta} - |z_n||} \geq \frac{1}{C+4}, \qquad (e^{i\theta} \in \mathbb{T}).$$

Hence, writing $z_n = r_n e^{i\theta_n}$, $n \geq 1$, we obtain

$$|B'(e^{i\theta})| \lesssim \sum_{n=1}^{\infty} \frac{1 - r_n}{|e^{i\theta} - r_n|^{2/\delta}}, \qquad (e^{i\theta} \in \mathbb{T}).$$

Then Corollary 5.10, with $\beta = 2/\delta$, reveals that

$$|B'(e^{i\theta})| \in \bigcap_{0 < p < \frac{\delta}{2 - \delta(1-\alpha)}} L^p(\mathbb{T}).$$

Now, apply Lemma 8.1.

The case $\delta = 1$ is of special interest. Also note that the case $\alpha = 1$ of the following special result is actually the content of Theorem 7.13.

Corollary 8.17 *Let $(z_n)_{n \geq 1}$ be a Blaschke sequence satisfying the condition*

$$\sum_{n=1}^{\infty} (1 - |z_n|)^\alpha < \infty$$

for some $0 < \alpha \leq 1$. Suppose, moreover, that the zeros are in a Stolz domain. Then

$$B' \in \bigcap_{0 < p < \frac{1}{1+\alpha}} H^p(\mathbb{D}).$$

Example 8.18 Fix $0 < \alpha < 1$. To ensure that Corollary 8.17 is sharp, we give a Blaschke sequence $(r_n)_{n \geq 1}$ satisfying the condition

$$\sum_{n=1}^{\infty} (1 - r_n)^\alpha < \infty,$$

and hence $B' \in \bigcap_{0 < p < \frac{1}{1+\alpha}} H^p(\mathbb{D})$, but

$$B' \notin H^{\frac{1}{1+\alpha}}(\mathbb{D}).$$

Suppose that $(\varepsilon_n)_{n \geq 1}$ is a positive decreasing sequence in the interval $(0, 1)$ such that $\varepsilon_n \asymp \varepsilon_{n+1}$ and

$$\sum_{n=1}^{\infty} \frac{1}{\log^{1+\alpha} 1/\varepsilon_n} < \infty.$$

For example, we may take $\varepsilon_n = 1/2^n$. Put

$$m_n = \left[\frac{1}{\varepsilon_n^\alpha \log^{1+\alpha} 1/\varepsilon_n} \right]$$

zeros in the interval $[1 - \varepsilon_n, 1 - \varepsilon_{n+1})$. We may put them at at the point $1 - \varepsilon_n$. Then

$$\sum_{n=1}^{\infty} (1 - r_n)^\alpha \asymp \sum_{n=1}^{\infty} m_n \varepsilon_n^\alpha \asymp \sum_{n=1}^{\infty} \frac{1}{\log^{1+\alpha} 1/\varepsilon_n} < \infty.$$

Suppose that $\varepsilon_{N+1} < |\theta| < \varepsilon_N$. Then

$$\begin{aligned}
f_B(\theta) &= \sum_{n=1}^{\infty} \frac{1 - r_n}{(1 - r_n)^2 + \theta^2} \\
&\asymp \sum_{n=1}^{\infty} \frac{m_n \varepsilon_n}{\varepsilon_n^2 + \theta^2} \\
&\geq \frac{m_N \varepsilon_N}{\varepsilon_N^2 + \theta^2} \geq \frac{m_N}{2\varepsilon_N} \\
&\asymp \frac{1}{(\varepsilon_n \log 1/\varepsilon_n)^{1+\alpha}} \\
&\asymp \frac{1}{(|\theta| \log 1/|\theta|)^{1+\alpha}}.
\end{aligned}$$

Hence, $f_B \notin L^{\frac{1}{1+\alpha}}(-\pi, \pi)$ which, by Lemma 8.1, ensures $B' \notin H^{\frac{1}{1+\alpha}}(\mathbb{D})$.

Example 8.19   Fix $0 < \alpha < 1$. We give a Blaschke sequence $(r_n)_{n \geq 1}$ such that $B' \in H^{\frac{1}{1+\alpha}}(\mathbb{D})$, and thus, by Theorem 8.7,

$$\sum_{n=1}^{\infty} (1 - r_n)^\alpha < \infty,$$

and yet

$$\sum_{n=1}^{\infty} (1 - r_n)^\beta = \infty$$

for all $\beta < \alpha$.

Fix $t > 1 + \alpha$. In the spirit of Example 8.18, suppose that $(\varepsilon_n)_{n \geq 1}$ is a positive decreasing sequence in the interval $(0, 1)$ such that $\varepsilon_n \asymp \varepsilon_{n+1}$ and

$$\sum_{n=1}^{\infty} \frac{1}{\log^t 1/\varepsilon_n} < \infty.$$

Moreover, we need

$$\sum_{n=1}^{N} \frac{1}{\varepsilon_n^{1+\alpha} \log^t 1/\varepsilon_n} \lesssim \frac{1}{\varepsilon_N^{1+\alpha} \log^{t'} 1/\varepsilon_N}$$

for any $t' < t$, and

$$\sum_{n=N+1}^{\infty} \frac{\varepsilon_n^{1-\alpha}}{\log^t 1/\varepsilon_n} \lesssim \frac{\varepsilon_N^{1-\alpha}}{\log^t 1/\varepsilon_N}.$$

However, still we may take $\varepsilon_n = 1/2^n$. Put

$$m_n = \left[\frac{1}{\varepsilon_n^{\alpha} \log^t 1/\varepsilon_n} \right]$$

zeros in the interval $[1 - \varepsilon_n, 1 - \varepsilon_{n+1})$. We may put them at the point $1 - \varepsilon_n$. Then, for each $\beta < \alpha$,

$$\sum_{n=1}^{\infty} (1 - r_n)^{\beta} \asymp \sum_{n=1}^{\infty} m_n \varepsilon_n^{\beta} \asymp \sum_{n=1}^{\infty} \frac{1}{\varepsilon_n^{\alpha-\beta} \log^t 1/\varepsilon_n} = \infty.$$

Suppose that $\varepsilon_{N+1} < |\theta| < \varepsilon_N$. Then, for each $t' < t$,

$$
\begin{aligned}
f_B(\theta) &= \sum_{n=1}^{\infty} \frac{1 - r_n}{(1 - r_n)^2 + \theta^2} \\
&\asymp \sum_{n=1}^{\infty} \frac{m_n \varepsilon_n}{\varepsilon_n^2 + \theta^2} \\
&\leq \sum_{n=1}^{N} \frac{m_n}{\varepsilon_n} + \sum_{n=N+1}^{\infty} \frac{m_n \varepsilon_n}{\theta^2} \\
&\asymp \sum_{n=1}^{N} \frac{1}{\varepsilon_n^{1+\alpha} \log^t 1/\varepsilon_n} + \frac{1}{\theta^2} \sum_{n=N+1}^{\infty} \frac{\varepsilon_n^{1-\alpha}}{\log^t 1/\varepsilon_n} \\
&\lesssim \frac{1}{\varepsilon_N^{1+\alpha} \log^{t'} 1/\varepsilon_N} + \frac{1}{\theta^2} \frac{\varepsilon_N^{1-\alpha}}{\log^t 1/\varepsilon_N} \\
&\asymp \frac{1}{|\theta|^{1+\alpha} \log^{t'} 1/|\theta|}.
\end{aligned}
$$

Now, it is enough to pick t' with $1 + \alpha < t' < t$. Hence, $f_B \in L^{\frac{1}{1+\alpha}}(-\pi, \pi)$ which, by Lemma 8.1, ensures $B' \in H^{\frac{1}{1+\alpha}}(\mathbb{D})$.

8.6 The Effect of Argument of Zeros on $\|B'\|_p$

In all sufficient conditions that we studied in this chapter, to arrive at the conclusion $B' \in H^p(\mathbb{D})$ we put a restriction on $|z_n|$. We did not pay enough attention to the effect of $\arg z_n$. The only exception was when the zeros are in a Stolz domain, where we saw that we can slightly improve our results. However, even in this case, we did not consider the precise value of $\arg z_n$. In this section, we study a result in which we consider both $|z_n|$ and $\arg z_n$. -

Theorem 8.20 (Ahern–Clark [2]) *Let* $z_n = r_n\, e^{i\theta_n}$, $n \geq 1$, *be a Blaschke sequence satisfying the condition*

$$\sum_{n=1}^{\infty} (1 - r_n)^{\alpha} < \infty$$

for some $0 < \alpha \leq 1$. *Let* E *be the closure of the set* $\{\, e^{i\theta_n} : n \geq 1 \,\}$. *Suppose that* E *has Lebesgue measure zero with complementary arcs of length* $(\ell_n)_{n \geq 1}$ *which fulfill*

$$\sum_{n=1}^{\infty} \ell_n^{\beta} < \infty$$

for some $0 < \beta < 1$. *Then*

$$B' \in H^{\frac{1-\beta}{1+\alpha}}(\mathbb{D}).$$

Remark: If the zeros are all on a radius, then this is precisely Corollary 8.17. More generally, if $\sum_{n=1}^{\infty} \ell_n^{\beta} < \infty$ is valid for all $\beta > 0$, then

$$B' \in \bigcap_{0 < p < \frac{1}{1+\alpha}} H^p(\mathbb{D}),$$

even if the zeros are not in a Stolz domain.

Proof. According to Theorem 4.15,

$$|B'(e^{i\theta})| = \sum_{n=1}^{\infty} \frac{1 - |z_n|^2}{|e^{i\theta} - z_n|^2}, \qquad (e^{i\theta} \in \mathbb{T}).$$

By Lemma 5.9,

$$\sum_{n=1}^{\infty} \frac{1 - |z_n|^2}{|e^{i\theta} - z_n|^2} \lesssim \frac{1}{d^{1+\alpha}(e^{i\theta})}, \qquad (e^{i\theta} \in \mathbb{T}),$$

where

$$d(e^{i\theta}) = \inf_{n \geq 1} |e^{i\theta} - e^{i\theta_n}|, \qquad (e^{i\theta} \in \mathbb{T}).$$

But, by (5.1),

$$\int_{\mathbb{T}} \frac{d\theta}{d^{1-\beta}(e^{i\theta})} \lesssim \sum_{n=1}^{\infty} \ell_n^{\beta} < \infty.$$

Therefore, we conclude

$$\sum_{n=1}^{\infty} \frac{1 - |z_n|^2}{|e^{i\theta} - z_n|^2} \in L^{\frac{1-\beta}{1+\alpha}}(\mathbb{T}).$$

Now, apply Lemma 8.1.

$$\int \frac{d\psi}{\sqrt{1 - k^2 \sin^2 \psi}} \geq \dots$$

This corresponds to

$$\dots$$

Chapter 9
B^p-Means of B'

9.1 A Sufficient Condition for $B' \in B^p(\mathbb{D})$

According to Theorem 6.1, we have

$$B' \in \bigcap_{0<p<\frac{1}{2}} B^p(\mathbb{D}) \tag{9.1}$$

for any Blaschke product B. Compare this result with Theorem 7.12. There is a Blaschke product B such that $B' \notin B^{\frac{1}{2}}(\mathbb{D})$. Hence, (9.1) is sharp and the Blaschke condition alone is not enough to conclude further results. Thus, we need to consider the Blaschke sequences which satisfy a stronger growth condition. Two such results are treated below.

Theorem 9.1 (Rudin [42]) *Let $(z_n)_{n\geq 1}$ be a Blaschke sequence satisfying the stronger condition*

$$\sum_{n=1}^{\infty} (1 - |z_n|) \log \frac{1}{(1 - |z_n|)} < \infty.$$

Then $B' \in B^{\frac{1}{2}}(\mathbb{D})$.

Proof. By (4.16),

$$|B'(z)| \leq \sum_{n=1}^{\infty} |b'_{z_n}(z)|, \qquad (z \in \mathbb{D}).$$

Hence, for any $0 < p < 1$,

$$\|B'\|_{B^p} \leq \sum_{n=1}^{\infty} \|b'_{z_n}\|_{B^p}.$$

J. Mashreghi, *Derivatives of Inner Functions*, Fields Institute Monographs 31, DOI 10.1007/978-1-4614-5611-7_9, © Springer Science+Business Media New York 2013

But, according to (4.24), with $p = \frac{1}{2}$, we have

$$\|B'\|_{B^{\frac{1}{2}}} \lesssim \sum_{n=1}^{\infty} (1 - |z_n|) \log \frac{1}{(1 - |z_n|)} < \infty.$$

Compare the following result with Theorem 8.4, in which the H^p-means of B' are studied. See also Lemma 6.8 and Theorem 6.9.

Theorem 9.2 (Protas [39]) *Let $(z_n)_{n \geq 1}$ be a Blaschke sequence satisfying the stronger condition*

$$\sum_{n=1}^{\infty} (1 - |z_n|)^{\alpha} < \infty$$

for some $0 < \alpha < 1$. Then $B' \in B^{\frac{1}{1+\alpha}}(\mathbb{D})$. In particular, if

$$\sum_{n=1}^{\infty} (1 - |z_n|)^{\frac{1}{2}} < \infty,$$

then $B' \in B^{\frac{2}{3}}(\mathbb{D})$.

Proof. By (4.16),

$$|B'(z)| \leq \sum_{n=1}^{\infty} |b'_{z_n}(z)|, \qquad (z \in \mathbb{D}).$$

Hence, for any $0 < p < 1$,

$$\|B'\|_{B^p} \leq \sum_{n=1}^{\infty} \|b'_{z_n}\|_{B^p}.$$

But, according to (4.24), with $p = \frac{1}{1+\alpha}$, we have

$$\|B'\|_{B^{\frac{1}{1+\alpha}}} \lesssim \sum_{n=1}^{\infty} (1 - |w|)^{\alpha} < \infty.$$

Example 9.3   Suppose that $0 < \alpha < \alpha_0 < 1$ are given. Take $\delta = 1 - \alpha$ and $\gamma \in (\frac{1}{\alpha_0}, \frac{1}{\alpha})$. Define

$$z_n = \left(1 - \frac{1}{n^{\gamma}}\right) \exp\left(\frac{i}{n^{\gamma \epsilon}}\right), \qquad (n \geq 1),$$

where $\epsilon \in \left(\frac{\gamma - 1}{\gamma}, \delta\right)$. Then

$$B' \notin B^{\frac{1}{1+\alpha}}(\mathbb{D}),$$

while the corresponding Blaschke sequence $(z_n)_{n \geq 1}$ satisfies

$$\sum_{n=1}^{\infty} (1 - |z_n|)^{\alpha_0} < \infty.$$

Fix $0 < \alpha < 1$. Then there is a Blaschke product with zeros $(r_n e^{i\theta_n})_{n \geq 1}$ such that

$$\sum_{n=1}^{\infty} (1 - r_n)^{\alpha} < \infty,$$

and thus $B' \in B^{\frac{1}{1+\alpha}}(\mathbb{D})$, but

$$B' \notin B^p(\mathbb{D})$$

for any $p > \frac{1}{1+\alpha}$. Hence, Theorem 9.2 is sharp. However, its converse is not true. There is a Blaschke product B such that $B' \in B^{\frac{1}{1+\alpha}}(\mathbb{D})$, and yet

$$\sum_{n=1}^{\infty} (1 - r_n)^{\alpha} = \infty.$$

9.2 What Does $B' \in B^p(\mathbb{D})$ Imply?

In Sect. 8.3, we saw that the assumption $B' \in H^p(\mathbb{D})$ puts a restriction on the rate of growth of $|z_n|$. None of those results were reversible. Nevertheless, a partial converse held. We study now similar results for $B^p(\mathbb{D})$ spaces in this section. Moreover, we emphasize that the following results are not reversible either.

Theorem 9.4 (Ahern–Clark [2]) *Let B be a Blaschke product formed with the Blaschke sequence $(z_n)_{n \geq 1}$. Let $\frac{2}{3} < p < 1$, and suppose that $B' \in B^p(\mathbb{D})$. Then*

$$\sum_{n=1}^{\infty} (1 - |z_n|)^{\alpha} < \infty$$

for all $\alpha > \frac{1-p}{2p-1}$.

Proof. By Theorem 6.3,

$$\|B'\|_{B^p} \gtrsim \int_0^{2\pi} \int_0^1 \left(1 - |B(re^{i\theta})|^2 \right) (1 - r)^{\frac{1}{p} - 3} \, dr d\theta,$$

and, by Theorem 3.5,

$$\frac{1 - |B(z)|^2}{1 - |z|^2} = \sum_{k=1}^{\infty} |B_k(z)|^2 \frac{1 - |z_k|^2}{|1 - \bar{z}_k z|^2},$$

where $B_1 = 1$ and

$$B_k(z) = \prod_{j=1}^{k-1} \frac{z_j - z}{1 - \bar{z}_j z}, \qquad (k \geq 2).$$

Hence,

$$\|B'\|_{B^p} \gtrsim \sum_{k=1}^{\infty} \int_0^{2\pi} \int_0^1 |B_k(re^{i\theta})|^2 \frac{(1 - |z_k|^2)(1 - r)^{\frac{1}{p} - 2}}{|1 - \bar{z}_k re^{i\theta}|^2} \, dr d\theta,$$

$$\geq \sum_{k=1}^{\infty} \int_0^{2\pi} \int_{\varrho_k}^1 |B_k(re^{i\theta})|^2 \frac{(1 - |z_k|^2)(1 - r)^{\frac{1}{p} - 2}}{|1 - \bar{z}_k re^{i\theta}|^2} \, dr d\theta,$$

where ϱ_k is any number in $[0, 1)$. But, the key to success is that there is a constant $c > 0$ and a proper choice of ϱ_k such that $\varrho_k > |z_k|$ and $|B_k(z)| \geq c$ on the annulus $\varrho_k \leq |z| < 1$ for all $k \geq 1$. Before introducing ϱ_k, let see what happens if this is true. If so, we would have

$$\|B'\|_{B^p} \gtrsim \sum_{k=1}^{\infty} \int_0^{2\pi} \int_{\varrho_k}^1 \frac{(1 - |z_k|^2)(1 - r)^{\frac{1}{p} - 2}}{|1 - \bar{z}_k re^{i\theta}|^2} \, dr d\theta$$

$$= \sum_{k=1}^{\infty} (1 - |z_k|^2) \int_{\varrho_k}^1 (1 - r)^{\frac{1}{p} - 2} \left(\int_0^{2\pi} \frac{1}{|1 - \bar{z}_k re^{i\theta}|^2} \, d\theta \right) dr$$

$$\asymp \sum_{k=1}^{\infty} (1 - |z_k|^2) \int_{\varrho_k}^1 \frac{(1 - r)^{\frac{1}{p} - 2}}{1 - r^2 |z_k|^2} \, dr$$

$$\gtrsim \sum_{k=1}^{\infty} (1 - |z_k|) \int_{\varrho_k}^1 \frac{(1 - r)^{\frac{1}{p} - 2}}{(1 - |z_k|) + (1 - r)} \, dr$$

$$= \sum_{k=1}^{\infty} (1 - |z_k|)^{\frac{1}{p} - 1} \int_0^{\frac{1 - \varrho_k}{1 - |z_k|}} \frac{t^{\frac{1}{p} - 2}}{1 + t} \, dt$$

$$\gtrsim \sum_{k=1}^{\infty} (1 - |z_k|)^{\frac{1}{p} - 1} \int_0^{\frac{1 - \varrho_k}{1 - |z_k|}} t^{\frac{1}{p} - 2} \, dt$$

$$\asymp \sum_{k=1}^{\infty} (1 - \varrho_k)^{\frac{1}{p} - 1}.$$

Hence,

$$\sum_{k=1}^{\infty} (1 - \varrho_k)^{\frac{1}{p} - 1} < \infty. \tag{9.2}$$

It remains to introduce ϱ_k. Let $\alpha \in (0,1]$ be such that

$$\sum_{k=1}^{\infty}(1 - |z_k|)^{\alpha} < \infty. \tag{9.3}$$

Note that, at least, this condition has to hold for $\alpha = 1$. Put $\rho_k = (1 - |z_k|)^{\alpha-1}$, $k \geq 1$. Since

$$-\log|z_k|^{\rho_k} = (1 - |z_k|)^{\alpha}\frac{-\log|z_k|}{1 - |z_k|} \sim (1 - |z_k|)^{\alpha},$$

the product $\prod_{k=1}^{\infty}|z_k|^{\rho_k}$ converges to a strictly positive number, say

$$c = \prod_{k=1}^{\infty}|z_k|^{\rho_k} > 0.$$

Hence, without loss of generality, we may assume that $|z_k|^{\rho_k} > \frac{1}{2}$ for all $k \geq 1$. Put

$$\varrho_k = \frac{|z_k| + |z_k|^{\rho_k}}{1 + |z_k|^{\rho_k+1}}, \qquad (k \geq 1).$$

Clearly, $|z_k| < \varrho_k < 1$, and

$$\frac{|z| - |z_k|}{1 - |z_k|\,|z|} \geq |z_k|^{\rho_k}, \qquad (k \geq 1),$$

on the annulus $\varrho_k \leq |z| < 1$. Moreover, ϱ_k is an increasing sequence (as usual we assume that $|z_k|$ is increasing). Hence, on the annulus $\varrho_k \leq |z| < 1$, we have

$$|B_k(z)| \geq \prod_{j=1}^{k-1}\left|\frac{z_j - z}{1 - \bar{z}_j\,z}\right|$$

$$\geq \prod_{j=1}^{k-1}\frac{|z| - |z_j|}{1 - |z_j|\,|z|}$$

$$\geq \prod_{j=1}^{k-1}|z_j|^{\rho_j} \geq \prod_{j=1}^{\infty}|z_j|^{\rho_j} = c.$$

To exploit (9.2), and finish the proof, we need to estimate $1 - \varrho_k$. By the definition of ϱ_k, we have

$$1 - \varrho_k = (1 - |z_k|) \frac{1 - |z_k|^{\rho_k}}{1 + |z_k|^{\rho_k + 1}}$$

$$\geq \frac{1}{2} (1 - |z_k|) (1 - |z_k|^{\rho_k})$$

$$\geq \frac{1}{4} \rho_k (1 - |z_k|)^2 = \frac{1}{4} (1 - |z_k|)^{\alpha+1}.$$

The last inequality comes from the mean value theorem. With function $f(x) = x^\rho$, on $[a, 1]$ with $a^\rho > \frac{1}{2}$, we have, for some $x \in (a, 1)$,

$$1 - a^\rho = \rho x^{\rho-1} (1 - a) \geq \rho a^\rho (1 - a) \geq \frac{1}{2} \rho (1 - a).$$

Therefore, by (9.2),

$$\sum_{k=1}^{\infty} (1 - |z_k|)^{\frac{(1+\alpha)(1-p)}{p}} < \infty. \tag{9.4}$$

In short, starting with the assumptions (9.3) and $B' \in B^p$, we deduced (9.4). Thus, define $\alpha_1 = 1$ and

$$\alpha_{n+1} = \frac{(1 + \alpha_n)(1 - p)}{p}, \qquad (n \geq 1).$$

Hence,

$$\sum_{k=1}^{\infty} (1 - |z_k|)^{\alpha_n} < \infty, \qquad (n \geq 1).$$

Finally, it is elementary so see that $(\alpha_n)_{n \geq 1}$ is a decreasing sequence and

$$\lim_{n \to \infty} \alpha_n = \frac{1 - p}{2p - 1}.$$

The assumption $\frac{2}{3} < p < 1$ was implicitly used here.

To be able to better compare Theorems 9.2 and 9.4, we can write them in the following ways. If $0 < \alpha < 1$, then

$$\sum_{n=1}^{\infty} (1 - |z_n|)^\alpha < \infty \implies B' \in B^{\frac{1}{1+\alpha}}(\mathbb{D}),$$

$$\sum_{n=1}^{\infty} (1 - |z_n|)^\alpha = \infty \implies B' \notin B^t(\mathbb{D}), \qquad \left(\frac{1+\alpha}{1+2\alpha} < t < 1 \right),$$

or equivalently, with $\frac{1}{2} < q < 1$ and $\frac{2}{3} < p < 1$,

$$B' \notin B^q(\mathbb{D}) \implies \sum_{n=1}^{\infty} (1 - |z_n|)^{\frac{1}{q}-1} = \infty,$$

$$B' \in B^p(\mathbb{D}) \implies \sum_{n=1}^{\infty} (1 - |z_n|)^{\beta} < \infty, \qquad \left(\beta > \frac{1-p}{2p-1} \right).$$

Clearly, none of the implications is reversible.

9.3 An Example of Blaschke Products

This section is a continuation of Sect. 8.4. We give one more interesting example of Blaschke products.

Example 9.5 The inclusion $H^p(\mathbb{D}) \subset B^p(\mathbb{D})$ created a hope that the results of Sects. 9.1 and 9.2 might be deducible from the corresponding results of Sects. 8.2 and 8.3. However, to ensure that this idea does not work, we construct a Blaschke product B such that $B' \in B^{\frac{1}{2}}(\mathbb{D})$ but $B' \notin H^{\frac{1}{2}}(\mathbb{D})$.

Let $(\varepsilon_n)_{n \geq 1}$ be any sequence in $(0,1)$ such that

$$\sum_{n=1}^{\infty} \varepsilon_n^{\frac{1}{2}} < \infty,$$

but

$$\sum_{n=1}^{\infty} \varepsilon_n^{\frac{1}{2}} \log \frac{1}{\varepsilon_n} = \infty.$$

For convenience, we assume that

$$\sum_{n=1}^{\infty} \varepsilon_n^{\frac{1}{2}} < \frac{\pi}{2}.$$

Put

$$\theta_n = \sum_{k=n}^{\infty} \varepsilon_k^{\frac{1}{2}},$$

and $r_n = 1 - \varepsilon_n$, $n \geq 1$. Note that θ_n is a positive decreasing sequence such that

$$\theta_n - \theta_{n+1} = \varepsilon_n^{\frac{1}{2}}, \qquad (n \geq 1),$$

and that

$$\lim_{n \to \infty} \theta_n = 0.$$

Therefore, $(r_n e^{i\theta_n})_{n \geq 1}$ is a Blaschke sequence tending to 1, as $n \longrightarrow \infty$, and, by Theorem 9.1, $B' \in B^{\frac{1}{2}}(\mathbb{D})$. To show that $B' \notin H^{\frac{1}{2}}(\mathbb{D})$, by Lemma 8.1, it is enough to prove that

$$f_B(\theta) = \sum_{n=1}^{\infty} \frac{1 - r_n}{(1 - r_n)^2 + (\theta - \theta_n)^2} \notin L^{\frac{1}{2}}(-\pi, \pi).$$

Since all the terms in the summation are positive, on the interval (θ_{n+1}, θ_n), we have

$$f_B(\theta) \geq \frac{1 - r_n}{(1 - r_n)^2 + (\theta - \theta_n)^2},$$

and thus

$$
\begin{aligned}
\int_{\theta_{n+1}}^{\theta_n} f_B^{\frac{1}{2}}(\theta)\, d\theta &\geq \int_{\theta_{n+1}}^{\theta_n} \frac{(1 - r_n)^{\frac{1}{2}}}{\left((1 - r_n)^2 + (\theta - \theta_n)^2 \right)^{\frac{1}{2}}}\, d\theta \\
&= \varepsilon_n^{\frac{1}{2}} \int_0^{\varepsilon_n^{\frac{1}{2}}} \frac{dt}{\left(\varepsilon_n^2 + t^2 \right)^{\frac{1}{2}}} \\
&= \varepsilon_n^{\frac{1}{2}} \int_0^{\varepsilon_n^{-\frac{1}{2}}} \frac{ds}{\left(1 + s^2 \right)^{\frac{1}{2}}} \\
&\geq \varepsilon_n^{\frac{1}{2}} \int_0^1 \frac{ds}{\sqrt{2}} + \varepsilon_n^{\frac{1}{2}} \int_1^{\varepsilon_n^{-\frac{1}{2}}} \frac{ds}{\sqrt{2}\, s} \\
&\geq \frac{1}{\sqrt{2}}\, \varepsilon_n^{\frac{1}{2}} + \frac{1}{2\sqrt{2}}\, \varepsilon_n^{\frac{1}{2}} \log \frac{1}{\varepsilon_n}.
\end{aligned}
$$

Hence, $f_B \notin L^{\frac{1}{2}}(-\pi, \pi)$.

9.4 The Effect of Argument of Zeros on $\|B'\|_{B_p}$

In all sufficient conditions that we studied in this chapter, to arrive at the conclusion $B' \in B^p(\mathbb{D})$ we put a restriction on $|z_n|$. We did not consider the effect of $\arg z_n$. In this section, we study a result in which we pay attention to $|z_n|$ and $\arg z_n$.

Theorem 9.6 (Ahern–Clark [4]) *Let $z_n = r_n\, e^{i\theta_n}$, $n \geq 1$, be a Blaschke sequence satisfying the condition*

$$\sum_{n=1}^{\infty} (1 - r_n)^\alpha < \infty$$

for some $0 < \alpha \leq 1$. Let E be the closure of the set $\{ e^{i\theta_n} : n \geq 1 \}$. Suppose that E has type β for some $0 < \beta \leq 1$. Then

$$\int_0^{2\pi} \left(1 - |B(re^{i\theta})| \right) d\theta \lesssim (1 - r)^q$$

for all $q < \frac{\beta}{1+\alpha}$. In particular, this estimation implies

$$B' \in \bigcap_{0<p<\frac{1+\alpha}{2+2\alpha-\beta}} B^p(\mathbb{D}).$$

Remark: We obtain a better result than Theorem 9.2 provided that $\beta > 1 - \alpha^2$.

Proof. By Theorem 3.5, we have

$$\frac{1-|B(z)|^2}{1-|z|^2} = \sum_{n=1}^{\infty} |B_n(z)|^2 \frac{1-|z_n|^2}{|1-\bar{z}_n z|^2}, \qquad (z \in \mathbb{D}).$$

Hence,

$$1 - |B(re^{i\theta})| \leq 8(1-r) \sum_{n=1}^{\infty} \frac{1-|z_n|^2}{|e^{i\theta}-z_n|^2}, \qquad (re^{i\theta} \in \mathbb{D}).$$

But, by Lemma 5.9,

$$\sum_{n=1}^{\infty} \frac{1-|z_n|^2}{|e^{i\theta}-z_n|^2} \leq \frac{8\sum_{n=1}^{\infty}(1-r_n)^\alpha}{d^{1+\alpha}(e^{i\theta})}, \qquad (e^{i\theta} \in \mathbb{T}),$$

where

$$d(e^{i\theta}) = \inf_{n\geq 1} |e^{i\theta} - e^{i\theta_n}|.$$

Therefore,

$$1 - |B(re^{i\theta})| \lesssim \min\left\{ 1, \frac{1-r}{d^{1+\alpha}(e^{i\theta})} \right\}, \qquad (re^{i\theta} \in \mathbb{D}). \qquad (9.5)$$

Note that our assumption on the type of E means that

$$\left| \{ e^{i\theta} \in \mathbb{T} : d(e^{i\theta}) \leq \varepsilon \} \right| \lesssim \varepsilon^\beta \qquad (9.6)$$

as $\varepsilon \longrightarrow 0$.

The rest of proof is similar to that of Theorem 6.14. however, for the sake of completeness, we give the details. Fix $q < \frac{\beta}{1+\alpha}$. Let $(\gamma_n)_{n\geq 0}$ be any decreasing sequence of real numbers and put

$$I_0 = \{ e^{i\theta} \in \mathbb{T} : d(e^{i\theta}) \leq 2(1-r)^{\gamma_0} \},$$

and, for $n \geq 1$,

$$I_n = \{ e^{i\theta} \in \mathbb{T} : 2(1-r)^{\gamma_{n-1}} < d(e^{i\theta}) \leq 2(1-r)^{\gamma_n} \}.$$

Hence, by (9.5) and (9.6),

$$\int_{I_0} (1 - |B(re^{i\theta})|)\, d\theta \leq |I_0| \lesssim (1-r)^{\beta\gamma_0}$$

and similarly, for $n \geq 1$,

$$\int_{I_n} (1 - |B(re^{i\theta})|)\, d\theta \lesssim \frac{1-r}{(1-r)^{(1+\alpha)\gamma_{n-1}}} \quad |I_n| \lesssim (1-r)^{1-(1+\alpha)\gamma_{n-1}+\beta\gamma_n}.$$

Thus, if $(\gamma_n)_{n \geq 0}$ is such that

$$\beta\gamma_0 = 1 - (1+\alpha)\gamma_{n-1} + \beta\gamma_n = q, \qquad (n \geq 1),$$

then we have

$$\int_{\cup_{n=0}^{N} I_n} (1 - |B(re^{i\theta})|)\, d\theta \lesssim N\,(1-r)^q. \qquad (9.7)$$

Two comments are in order. First, the sequence $(\gamma_n)_{n \geq 0}$ defined by

$$\gamma_0 = \frac{q}{\beta} \qquad \text{and} \qquad \gamma_n = \frac{(1+\alpha)\gamma_{n-1} + q - 1}{\beta}, \qquad (n \geq 1),$$

is decreasing. This can be easily verified by relations

$$\gamma_0 - \gamma_1 = \frac{\beta - (1+\alpha)q}{\beta^2}$$

and

$$\gamma_n - \gamma_{n-1} = \frac{(1+\alpha)(\gamma_{n-1} - \gamma_{n-2})}{\beta}, \qquad (n \geq 1).$$

Second, for $n \geq 1$, the unique solution is also given by

$$\left(\frac{\beta}{1+\alpha}\right)^n \gamma_n = \frac{q}{\beta} - \frac{1-q}{1+\alpha}\left((1 + (\frac{\beta}{1+\alpha}) + (\frac{\beta}{1+\alpha})^2 + \cdots + (\frac{\beta}{1+\alpha})^{n-1}\right).$$

As $n \longrightarrow \infty$, the right side converges to

$$\frac{q}{\beta} - \frac{1-q}{1+\alpha} \times \frac{1}{1 - \frac{\beta}{1+\alpha}} = \frac{(1+\alpha)q - \beta}{\beta(1+\alpha-\beta))} < 0.$$

Hence, there is an N_0 such that $\gamma_{N_0} < 0$. Therefore, $\mathbb{T} = \cup_{n=0}^{N_0} I_n$, and by (9.7), we get

$$\int_{\mathbb{T}} (1 - |B(re^{i\theta})|)\, d\theta \lesssim (1-r)^q.$$

By Corollary 6.6, this estimation implies $B' \in B^p(\mathbb{D})$, for any $0 < p < \frac{1}{2-q}$. Thus, $B' \in B^p(\mathbb{D})$, for all $0 < p < \frac{1+\alpha}{2+2\alpha-\beta}$.

Corollary 9.7 *Let $z_n = r_n\, e^{i\theta_n}$, $n \geq 1$, be a Blaschke sequence satisfying the condition*

$$\sum_{n=1}^{\infty}(1 - r_n)^{\alpha} < \infty$$

for some $0 < \alpha \leq 1$. Let E be the closure of the set $\{\, e^{i\theta_n} : n \geq 1 \,\}$. Suppose that E has type β for all $\beta < 1$. Then

$$B' \in \bigcap_{0<p<\frac{1+\alpha}{1+2\alpha}} B^p(\mathbb{D}).$$

Chapter 10
The Growth of Integral Means of B'

In all the preceding chapters, we studied various conditions under which some integral means of B' were uniformly bounded. In this chapter, on the contrary, we assume that the integral means are not bounded and, in fact, they tend to infinity as a parameter varies. This parameter is usually the radius r which tends to 1. Our goal is to study the rate of growth of integral means as $r \longrightarrow 1$.

10.1 An Estimation Lemma

Let B be a Blaschke product with zeros at $(z_n)_{n \geq 1}$. Then,

$$
\int_0^{2\pi} |B'(re^{i\theta})| \, d\theta \leq \sum_{n=1}^{\infty} (1 - |z_n|^2) \int_0^{2\pi} \frac{d\theta}{|1 - \bar{z}_n \, re^{i\theta}|^2}
$$

$$
= \sum_{n=1}^{\infty} (1 - |z_n|^2) \frac{2\pi}{(1 - |z_n|^2 r^2)}
$$

$$
\leq \frac{4\pi \sum_{n=1}^{\infty} (1 - |z_n|)}{(1 - r)},
$$

which implies

$$
\int_0^{2\pi} |B'(re^{i\theta})| \, d\theta = \frac{o(1)}{1 - r}, \qquad (r \longrightarrow 1). \tag{10.1}
$$

However, assuming stronger restrictions on the rate of increase of the zeros of B give us more precise estimates about the rate of increase of integral means of B_r' as $r \longrightarrow 1$. The most common restriction is

J. Mashreghi, *Derivatives of Inner Functions*, Fields Institute Monographs 31,   157
DOI 10.1007/978-1-4614-5611-7_10, © Springer Science+Business Media New York 2013

$$\sum_{n=1}^{\infty} (1 - |z_n|)^{\alpha} < \infty \tag{10.2}$$

for some $\alpha \in (0,1)$. We consider the more general assumption

$$\sum_{n=1}^{\infty} h(1 - |z_n|) < \infty, \tag{10.3}$$

where h is a positive continuous function satisfying certain smoothness conditions. In particular, we are interested in the functions

$$h(t) = t^{\alpha} \, (\log 1/t)^{\alpha_1} \, (\log_2 1/t)^{\alpha_2} \, \cdots \, (\log_n 1/t)^{\alpha_n}, \tag{10.4}$$

where $\alpha \in (0,1)$, $\alpha_1, \alpha_2, \cdots, \alpha_n \in \mathbb{R}$, and $\log_n = \log \log \cdots \log$ (n times).

In the following we assume that h is a continuous positive function defined on the interval $(0,1)$ with

$$\lim_{t \to 0+} h(t) = 0.$$

Our prototype is the one given in (10.4). The following lemma has simple assumptions and also a very simple proof. However, it has some interesting applications.

Lemma 10.1 *Let $(r_n)_{n \geq 1}$ be a sequence in the interval $(0,1)$ such that*

$$\sum_{n=1}^{\infty} h(1 - r_n) < \infty.$$

Let $p > 0$ and $q > 0$ be such that $h(t)/t^p$ is decreasing and $h(t)/t^{p-q}$ is increasing on $(0,1)$. Then,

$$\sum_{n=1}^{\infty} \frac{(1 - r_n)^p}{(1 - rr_n)^q} = \frac{O(1)}{(1 - r)^{q-p} \, h(1 - r)}$$

as $r \longrightarrow 1$. Moreover, if

$$\lim_{t \to 0+} \frac{h(t)}{t^{p-q}} = 0,$$

then

$$\sum_{n=1}^{\infty} \frac{(1 - r_n)^p}{(1 - rr_n)^q} = \frac{o(1)}{(1 - r)^{q-p} \, h(1 - r)}.$$

Proof. We have

$$\frac{(1 - r_n)^p}{(1 - rr_n)^q} = \left(\frac{(1 - r_n)^p}{h(1 - r_n)} \frac{h(1 - rr_n)}{(1 - rr_n)^p} \right) \left(\frac{h(1 - r_n)}{(1 - rr_n)^{q-p} \, h(1 - rr_n)} \right).$$

By assumption

$$\frac{h(1 - rr_n)}{(1 - rr_n)^p} \leq \frac{h(1 - r_n)}{(1 - r_n)^p},$$

and

$$(1 - rr_n)^{q-p} h(1 - rr_n) \geq (1 - r)^{q-p} h(1 - r).$$

Thus, for all $n \geq 1$,

$$\frac{(1 - r_n)^p}{(1 - rr_n)^q} \leq \frac{h(1 - r_n)}{(1 - r)^{q-p} h(1 - r)}. \tag{10.5}$$

Given $\varepsilon > 0$, fix N such that

$$\sum_{n=N+1}^{\infty} h(1 - r_n) < \varepsilon.$$

Hence, by (10.5),

$$\sum_{n=1}^{\infty} \frac{(1 - r_n)^p}{(1 - rr_n)^q} = \sum_{n=1}^{N} \frac{(1 - r_n)^p}{(1 - rr_n)^q} + \sum_{n=N+1}^{\infty} \frac{(1 - r_n)^p}{(1 - rr_n)^q}$$

$$\leq \sum_{n=1}^{N} (1 - r_n)^{p-q} + \frac{\sum_{n=N+1}^{\infty} h(1 - r_n)}{(1 - r)^{q-p} h(1 - r)}$$

$$\leq C_N + \frac{\varepsilon}{(1 - r)^{q-p} h(1 - r)}.$$

This inequality implies both assertions of the lemma.

Lemma 10.1 is still valid if instead of "φ being increasing", we assume that there is a constant $C > 0$ such that $\varphi(x) \leq C\varphi(y)$ whenever $x \leq y$. The same comment applies for the theorems of remaining sections.

10.2 H^p-Means of the First Derivative

In this section we apply Lemma 10.1 to obtain a general estimate for the H^p-means of the first derivative of a Blaschke product.

Theorem 10.2 (Fricain–Mashreghi [21]) *Let B be the Blaschke product formed with zeros $z_n = r_n e^{i\theta_n}$, $n \geq 1$, satisfying*

$$\sum_{n=1}^{\infty} h(1 - r_n) < \infty$$

for a positive continuous function h. Suppose that there is $q \in (1/2, 1]$ such that $h(t)/t^q$ is decreasing and $h(t)/t^{1-q}$ is increasing on $(0, 1)$. Then, for any $p \geq q$,

$$\int_0^{2\pi} |B'(re^{i\theta})|^p \, d\theta = \frac{O(1)}{(1-r)^{p-1} h(1-r)}, \qquad (r \longrightarrow 1).$$

Moreover, if $\lim_{t \to 0} h(t)/t^{1-q} = 0$, then $O(1)$ can be replaced by $o(1)$.

Proof. Since $q \leq 1$, (4.16) implies

$$|B'(re^{i\theta})|^q \leq \sum_{n=1}^{\infty} \frac{(1-r_n^2)^q}{|1 - rr_n e^{i(\theta-\theta_n)}|^{2q}}.$$

Hence

$$\int_0^{2\pi} |B'(re^{i\theta})|^q \, d\theta \leq C \sum_{n=1}^{\infty} \frac{(1-r_n)^q}{(1-rr_n)^{2q-1}}. \qquad (10.6)$$

(Here we used the assumption $2q > 1$.) Therefore, by Lemma 10.1,

$$\int_0^{2\pi} |B'(re^{i\theta})|^q \, d\theta \leq \frac{C}{(1-r)^{q-1} h(1-r)}.$$

Any H^∞-function is in the Bloch space. Hence, for any $p \geq q$,

$$\int_0^{2\pi} |B'(re^{i\theta})|^p \, d\theta \leq \frac{1}{(1-r)^{p-q}} \int_0^{2\pi} |B'(re^{i\theta})|^q \, d\theta \leq \frac{C}{(1-r)^{p-1} h(1-r)}.$$

Finally, as $r \longrightarrow 1$, Lemma 10.1 also assures that the constant C can be replaced by any small positive constant if $\lim_{t \to 0} h(t)/t^{1-q} = 0$.

Now, we can apply Theorem 10.2 for the special function h given in (10.4).

Case I: If

$$\sum_{n=1}^{\infty} (1-r_n)^\alpha (\log \frac{1}{1-r_n})^{\alpha_1} \cdots (\log_m \frac{1}{1-r_n})^{\alpha_m} < \infty,$$

then, for any

$$p > \max\{\alpha, 1 - \alpha\},$$

we have

$$\int_0^{2\pi} |B'(re^{i\theta})|^p \, d\theta = \frac{o(1)}{(1-r)^{\alpha+p-1}(\log \frac{1}{1-r})^{\alpha_1} \cdots (\log_m \frac{1}{1-r})^{\alpha_m}},$$

as $r \longrightarrow 1$. In particular, if

$$\sum_{n=1}^{\infty} (1 - r_n)^{\alpha} < \infty,$$

with $\alpha \in (0, 1/2)$, then, for any $p > 1 - \alpha$,

$$\int_0^{2\pi} |B'(re^{i\theta})|^p \, d\theta = \frac{o(1)}{(1-r)^{p+\alpha-1}}, \qquad (r \longrightarrow 1).$$

Moreover, if $\alpha \in [1/2, 1)$, the last estimate still holds for any $p > \alpha$.

Case II: If

$$\sum_{n=1}^{\infty} (1 - r_n)^{\alpha} (\log_k \frac{1}{1 - r_n})^{\alpha_k} \cdots (\log_m \frac{1}{1 - r_n})^{\alpha_m} < \infty,$$

with $\alpha \in (0, 1/2)$, $\alpha_k < 0$ and $\alpha_{k+1}, \cdots, \alpha_n \in \mathbb{R}$, then,

$$\int_0^{2\pi} |B'(re^{i\theta})|^{1-\alpha} \, d\theta = \frac{o(1)}{(\log_k \frac{1}{1-r})^{\alpha_k} \cdots (\log_m \frac{1}{1-r})^{\alpha_m}}, \qquad (r \longrightarrow 1).$$

But, if

$$\sum_{n=1}^{\infty} (1 - r_n)^{\alpha} < \infty,$$

with $\alpha \in (0, 1/2)$, then

$$\int_0^{2\pi} |B'(re^{i\theta})|^{1-\alpha} \, d\theta = O(1), \qquad (r \longrightarrow 1),$$

i.e. $B' \in H^{1-\alpha}$, which is a Protas' result [39].

Case III: If

$$\sum_{n=1}^{\infty} (1 - r_n)^{\alpha} (\log_k \frac{1}{1 - r_n})^{\alpha_k} \cdots (\log_m \frac{1}{1 - r_n})^{\alpha_m} < \infty,$$

with $\alpha \in (1/2, 1)$, $\alpha_k > 0$ and $\alpha_{k+1}, \cdots, \alpha_n \in \mathbb{R}$, then,

$$\int_0^{2\pi} |B'(re^{i\theta})|^{\alpha} \, d\theta = \frac{o(1)}{(1-r)^{2\alpha-1}(\log \frac{1}{1-r})^{\alpha_1} \cdots (\log_m \frac{1}{1-r})^{\alpha_m}}, \qquad (r \longrightarrow 1).$$

However, if

$$\sum_{n=1}^{\infty} (1 - r_n)^{\alpha} < \infty,$$

with $\alpha \in (1/2, 1)$, then we still have

$$\int_0^{2\pi} |B'(re^{i\theta})|^\alpha \, d\theta = \frac{o(1)}{(1-r)^{2\alpha-1}}, \qquad (r \longrightarrow 1).$$

Theorem 10.2 has been further generalized by J. Mashreghi and M. Sha-bankhah [34]. Briefly speaking, they showed that similar estimates hold for the logarithmic derivative of B.

10.3 H^p-Means of Higher Derivatives

Let ℓ be a positive integer, and let $\frac{1}{\ell+1} < p \leq \frac{1}{\ell}$. Then straightforward calculation leads to the estimation

$$\int_0^{2\pi} |B^{(\ell)}(re^{i\theta})|^p \, d\theta \leq C(p,\ell) \sum_{n=1}^{\infty} \frac{(1-r_n)^p}{(1-rr_n)^{(\ell+1)p-1}}, \qquad (10.7)$$

which is a generalization of (10.6). This observation along with Lemma 10.1 enable us to generalize the results of the preceding section for higher deriva-tives of a Blaschke product. We omit the proof, since it is similar to the proof of Theorem 10.2.

Theorem 10.3 (Fricain–Mashreghi [21]) *Let B be the Blaschke product formed with zeros $z_n = r_n e^{i\theta_n}$, $n \geq 1$, satisfying*

$$\sum_{n=1}^{\infty} h(1 - r_n) < \infty$$

for a positive continuous function h. Suppose that there is $q \in (1/(\ell+1), 1/\ell]$ such that $h(t)/t^q$ is decreasing and $h(t)/t^{1-\ell q}$ is increasing on $(0,1)$. Then, for any $p \geq q$,

$$\int_0^{2\pi} |B^{(\ell)}(re^{i\theta})|^p \, d\theta = \frac{O(1)}{(1-r)^{\ell p-1} h(1-r)}, \qquad (r \longrightarrow 1).$$

Moreover, if $\lim_{t \to 0} h(t)/t^{1-\ell q} = 0$, then $O(1)$ can be replaced by $o(1)$.

Now, we can apply Theorem 10.3 for the special function h given in (10.4).

Case I: If

$$\sum_{n=1}^{\infty} (1-r_n)^\alpha (\log \frac{1}{1-r_n})^{\alpha_1} \cdots (\log_m \frac{1}{1-r_n})^{\alpha_m} < \infty,$$

then, for any
$$p > \max\{\alpha, (1-\alpha)/\ell\},$$
we have
$$\int_0^{2\pi} |B^{(\ell)}(re^{i\theta})|^p \, d\theta = \frac{o(1)}{(1-r)^{\alpha+\ell p-1}(\log\frac{1}{1-r})^{\alpha_1} \cdots (\log_m \frac{1}{1-r})^{\alpha_m}},$$
as $r \longrightarrow 1$. In particular, if
$$\sum_{n=1}^{\infty} (1-r_n)^{\alpha} < \infty,$$
with $\alpha \in (0, 1/(\ell+1))$, then, for any $p > (1-\alpha)/\ell$,
$$\int_0^{2\pi} |B^{(\ell)}(re^{i\theta})|^p \, d\theta = \frac{o(1)}{(1-r)^{\ell p+\alpha-1}}, \qquad (r \longrightarrow 1).$$

Moreover, if $\alpha \in [1/(\ell+1), 1)$, the last estimate still holds for any $p > \alpha$.

Case II: If
$$\sum_{n=1}^{\infty} (1-r_n)^{\alpha} (\log_k \frac{1}{1-r_n})^{\alpha_k} \cdots (\log_m \frac{1}{1-r_n})^{\alpha_m} < \infty,$$
with $\alpha \in (0, 1/(1+\ell))$, $\alpha_k < 0$ and $\alpha_{k+1}, \cdots, \alpha_n \in \mathbb{R}$, then,
$$\int_0^{2\pi} |B^{(\ell)}(re^{i\theta})|^{(1-\alpha)/\ell} \, d\theta = \frac{o(1)}{(\log_k \frac{1}{1-r})^{\alpha_k} \cdots (\log_m \frac{1}{1-r})^{\alpha_m}}, \qquad (r \longrightarrow 1).$$

But, if
$$\sum_{n=1}^{\infty} (1-r_n)^{\alpha} < \infty,$$
with $\alpha \in (0, 1/(1+\ell))$, then
$$\int_0^{2\pi} |B^{(\ell)}(re^{i\theta})|^{(1-\alpha)/\ell} \, d\theta = O(1), \qquad (r \longrightarrow 1),$$
i.e. $B^{(\ell)} \in H^{(1-\alpha)/\ell}$ which is a Linden's result [30].

Case III: If
$$\sum_{n=1}^{\infty} (1-r_n)^{\alpha} (\log_k \frac{1}{1-r_n})^{\alpha_k} \cdots (\log_m \frac{1}{1-r_n})^{\alpha_m} < \infty,$$

with $\alpha \in (1/(1+\ell), 1)$, $\alpha_k > 0$ and $\alpha_{k+1}, \cdots, \alpha_n \in \mathbb{R}$, then

$$\int_0^{2\pi} |B^{(\ell)}(re^{i\theta})|^\alpha \, d\theta = \frac{o(1)}{(1-r)^{(\ell+1)\alpha-1}(\log_k \frac{1}{1-r})^{\alpha_k} \cdots (\log_m \frac{1}{1-r})^{\alpha_m}},$$

as $r \longrightarrow 1$. However, if

$$\sum_{n=1}^{\infty} (1 - r_n)^\alpha < \infty,$$

with $\alpha \in (1/(\ell+1), 1)$, then we still have

$$\int_0^{2\pi} |B^{(\ell)}(re^{i\theta})|^\alpha \, d\theta = \frac{o(1)}{(1-r)^{(\ell+1)\alpha-1}}, \qquad (r \longrightarrow 1).$$

10.4 A_γ^p-Means of the First Derivative

In this section, we apply Lemma 10.1 to obtain a general estimate for the A_γ^p-means of the first derivative of a Blaschke product.

Theorem 10.4 (Fricain–Mashreghi [21]) *Let B be the Blaschke product formed with zeros $z_n = r_n e^{i\theta_n}$ satisfying*

$$\sum_{n=1}^{\infty} h(1 - r_n) < \infty$$

for a positive continuous function h. Let $\gamma \in (-1, 0)$. Suppose that there is $q \in (1 + \gamma/2, 1]$ such that $h(t)/t^q$ is decreasing and $h(t)/t^{2+\gamma-q}$ is increasing on $(0, 1)$. Then, for any $p \geq q$,

$$\int_0^1 \int_0^{2\pi} |B'(r\rho e^{i\theta})|^p \, \rho(1-\rho^2)^\gamma d\rho \, d\theta = \frac{O(1)}{(1-r)^{p-\gamma-2} h(1-r)}, \qquad (r \longrightarrow 1).$$

Moreover, if $\lim_{t\to 0} h(t)/t^{2+\gamma-q} = 0$, then $O(1)$ can be replaced by $o(1)$.

Remark. For the simplicity of notation, for the rest of this section we write I_r for the last integral, i.e.

$$I_r = \int_0^1 \int_0^{2\pi} |B'(r\rho e^{i\theta})|^p \, \rho(1-\rho^2)^\gamma d\rho \, d\theta.$$

Surely, the integral also depends on the other parameters. However, its dependence to r is important for us.

Proof. We saw that

$$|B'(r\rho e^{i\theta})|^q \leq \sum_{n=1}^{\infty} \frac{(1-r_n^2)^q}{|1-rr_n\rho e^{i(\theta-\theta_n)}|^{2q}}.$$

Hence,

$$I_r \leq C \sum_{n=1}^{\infty} \frac{(1-r_n)^q}{(1-rr_n)^{2q-\gamma-2}}. \tag{10.8}$$

(Here we used the assumption $2q - \gamma - 2 > 0$.) Therefore, by Lemma 10.1,

$$I_r \leq \frac{C}{(1-r)^{q-\gamma-2}\, h(1-r)}.$$

Thus, for any $p \geq q$,

$$I_r \leq \frac{1}{(1-r)^{p-q}} \int_0^1 \int_0^{2\pi} |B'(r\rho e^{i\theta})|^q \; \rho(1-\rho^2)^\gamma d\rho\, d\theta$$

$$\leq \frac{C}{(1-r)^{p-\gamma-2}\, h(1-r)}.$$

Finally, as $r \longrightarrow 1$, Lemma 10.1 also ensures that the constant C can be replaced by any small positive constant if $\lim_{t\to 0} h(t)/t^{2+\gamma-q} = 0$.

Now, we can apply Theorem 10.4 for the special function h given in (10.4).

Case I: If

$$\sum_{n=1}^{\infty}(1-r_n)^\alpha (\log\frac{1}{1-r_n})^{\alpha_1} \cdots (\log_m \frac{1}{1-r_n})^{\alpha_m} < \infty,$$

and if $\gamma \in (-1, \alpha-1)$, then, for any

$$p > \max\{\, \alpha,\, 2+\gamma-\alpha,\, 1+\gamma/2\,\},$$

we have

$$I_r = \frac{o(1)}{(1-r)^{\alpha+p-\gamma-2} (\log\frac{1}{1-r})^{\alpha_1} \cdots (\log_m \frac{1}{1-r})^{\alpha_m}},$$

as $r \longrightarrow 1$. In particular, if

$$\sum_{n=1}^{\infty}(1-r_n)^\alpha < \infty,$$

then

$$I_r = \frac{o(1)}{(1-r)^{p+\alpha-\gamma-2}}.$$

Case II: If

$$\sum_{n=1}^{\infty}(1-r_n)^{\alpha}(\log_k \frac{1}{1-r_n})^{\alpha_k}\cdots(\log_m \frac{1}{1-r_n})^{\alpha_m} < \infty,$$

with $\alpha_k < 0$, then, for any $p \geq 1$,

$$I_r = \frac{o(1)}{(1-r)^{p-1}(\log_k \frac{1}{1-r})^{\alpha_k}\cdots(\log_m \frac{1}{1-r})^{\alpha_m}},$$

as $r \longrightarrow 1$.

Case III: If

$$\sum_{n=1}^{\infty}(1-r_n)^{\alpha} < \infty,$$

then, for any $p \geq 1$,

$$I_r = \frac{O(1)}{(1-r)^{p-1}}, \qquad (r \longrightarrow 1).$$

In particular, for $p = 1$,

$$I_r = O(1), \qquad (r \longrightarrow 1),$$

which is a Protas' result [39].

Some other cases can also be considered here. But, since they are immediate consequences of Theorem 10.4, we do not proceed further. Moreover, using similar techniques, one can obtain estimates for the A_γ^p means of the higher derivatives for a Blaschke product satisfying the hypothesis of Theorem 10.4.

References

1. Ahern P (1979) On a theorem of Hayman concerning the derivative of a function of bounded characteristic. Pacific J Math 83(2):297–301
2. Ahern P, Clark D (1971) Radial nth derivatives of Blaschke products. Math Scand 28:189–201
3. Ahern P, Clark D (1974) On inner functions with H^p-derivative. Michigan Math J 21:115–127
4. Ahern P, Clark D (1976) On inner functions with B^p derivative. Michigan Math J 23(2):107–118
5. Allen H, Belna C (1972) Singular inner functions with derivative in B^p. Michigan Math J 19:185–188
6. Belna C, Muckenhoupt B (1977) The derivative of the atomic function is not in $B^{2/3}$. Proc Am Math Soc 63(1):129–130
7. Belna C, Colwell P, Piranian G (1985) The radial behavior of Blaschke products. Proc Amer Math Soc 93(2):267–271
8. Beurling A (1948) On two problems concerning linear transformations in Hilbert space. Acta Math 81:239–255
9. Blaschke W (1915) Eine erweiterung des satzes von vitali über folgen analytischer funktionen. Leipzig Ber 67:194–200
10. Bourgain J (1993) On the radial variation of bounded analytic functions on the disc. Duke Math J 69(3):671–682
11. Carathéodory C (1929) Über die winkelderivierten von beschränkten analytischen funktionen. Sitzunber Preuss Akad Wiss 32:39–52
12. Cargo G (1961) The radial images of Blaschke products. J London Math Soc 36: 424–430
13. Caughran J, Shields A (1969) Singular inner factors of analytic functions. Michigan Math J 16:409–410
14. Cohn W (1983) On the H^p classes of derivatives of functions orthogonal to invariant subspaces. Michigan Math J 30(2):221–229
15. Collingwood EF, Lohwater AJ (1966) The theory of cluster sets. Cambridge tracts in mathematics and mathematical physics, No 56. Cambridge University Press, Cambridge
16. Colwell P (1985) Blaschke products. University of Michigan Press, Ann Arbor. Bounded analytic functions
17. Cullen M (1971) Derivatives of singular inner functions. Michigan Math J 18:283–287
18. Duren P, Schuster A (2004) Bergman spaces, volume 100 of mathematical surveys and monographs. American Mathematical Society, Providence
19. Duren P, Romberg B, Shields A (1969) Linear functionals on H^p spaces with $0 < p < 1$. J Reine Angew Math 238:32–60

J. Mashreghi, *Derivatives of Inner Functions*, Fields Institute Monographs 31, 167
DOI 10.1007/978-1-4614-5611-7, © Springer Science+Business Media New York 2013

20. Fatou P (1906) Séries trigonométriques et séries de Taylor. Acta Math 30(1):335–400
21. Fricain E, Mashreghi J (2008) Integral means of the derivatives of Blaschke products. Glasg Math J 50(2):233–249
22. Frostman O (1935) Potentiel d'équilibre et capacité des ensembles avec quelques applications à la théorie des fonctions. Meddel Lund Univ Mat Sem 3
23. Frostman O (1942) Sur les produits de Blaschke. Kungl. Fysiografiska Sällskapets i Lund Förhandlingar [Proc Roy Physiog Soc Lund] 12(15):169–182
24. Girela D, Peláez J, Vukotić D (2007) Integrability of the derivative of a Blaschke product. Proc Edinb Math Soc (2), 50(3):673–687
25. Hardy G, Littlewood J (1932) Some properties of fractional integrals. II. Math Z 34(1):403–439
26. Hedenmalm H, Korenblum B, Zhu K (2000) Theory of Bergman spaces, volume 199 of graduate texts in mathematics. Springer, New York
27. Heins M (1951) A residue theorem for finite Blaschke products. Proc Amer Math Soc 2:622–624
28. Herglotz G (1911) Über potenzreihen mit positivem, reellen teil in einheitskreis. S-B Sächs Akad Wiss Leipzig Math-Natur Kl 63:501–511
29. Julia G (1920) Extension nouvelle d'un lemme de Schwarz. Acta Math 42(1):349–355
30. Linden C (1976) H^p-derivatives of Blaschke products. Michigan Math J 23(1):43–51
31. Lucas F (1874) Propriétés géométriques des fractionnes rationnelles. CR Acad Sci Paris 77:631–633
32. Mashreghi J (2002) Expanding a finite Blaschke product. Complex Var Theory Appl 47(3):255–258
33. Mashreghi J (2009) Representation theorems in Hardy spaces, volume 74 of london mathematical society student texts. Cambridge University Press, Cambridge
34. Mashreghi J, Shabankhah M (2009) Integral means of the logarithmic derivative of Blaschke products. Comput Methods Funct Theory 9(2):421–433
35. Nevanlinna R (1929) Über beschränkte analytische funcktionen. Ann Acad Sci Fennicae A 32(7):1–75
36. Plessner A (1923) Zur theorie der konjugierten trigonometrischen reihen. Mitt Math Sem Giessen 10:1–36
37. Privalov I (1918) Intégral de cauchy. Bulletin de l'Université, à Saratov
38. Privalov I (1924) Sur certaines propriétés métriques des fonctions analytiques. J de l'École Polytech 24:77–112
39. Protas D (1973) Blaschke products with derivative in H^p and B^p. Michigan Math J 20:393–396
40. Riesz F (1923) Über die Randwerte einer analytischen Funktion. Math Z 18(1):87–95
41. Riesz M (1931) Sur certaines inégalités dans la théorie des fonctions avec quelques remarques sur les géometries non-euclidiennes. Kungl Fysiogr Sällsk i Lund 1(4): 18–38
42. Rudin W (1955) The radial variation of analytic functions. Duke Math J 22:235–242
43. Seidel W (1934) On the distribution of values of bounded analytic functions. Trans Am Math Soc 36(1):201–226
44. Tsuji M (1959) Potential theory in modern function theory. Maruzen, Tokyo
45. Vukotić D (2003) The isoperimetric inequality and a theorem of Hardy and Littlewood. Am Math Mon 110(6):532–536
46. Walsh JL (1939) Note on the location of zeros of the derivative of a rational function whose zeros and poles are symmetric in a circle. Bull Am Math Soc 45(6):462–470

Index

J. Mashreghi, *Derivatives of Inner Functions*, Fields Institute Monographs 31, DOI 10.1007/978-1-4614-5611-7, © Springer Science+Business Media New York 2013